AF792885

THE NORMATIVITY OF THE NATURAL

Philosophical Studies in Contemporary Culture

VOLUME 16

Senior Editor

H. Tristram Engelhardt, Jr., *Department of Philosophy, Rice University, and Baylor College of Medicine, Houston, Texas*

Editor

Mark J. Cherry, *Department of Philosophy, St. Edward's University, Austin, Texas*

Assistant Editor

Lisa Rasmussen, *Department of Philosophy, University of North Carolina, Charlotte, North Carolina*

Editorial Board

Stanley Hauerwas, *Divinity School, Duke University, Durham, North Carolina*

Maureen Kelley, *Department of Pediatric Bioethics, University of Washington, Seattle, Washington*

Terry Pinkard, *Department of Philosophy, Georgetown University, Washington, D.C.*

C. Griffin Trotter, *Department of Health Care Ethics & Emergency Medicine, Saint Louis University, St. Louis, Missouri*

Kevin Wm. Wildes, S.J., *President, Loyola University, New Orleans, Louisiana*

For further volumes:
http://www.springer.com/series/6446

THE NORMATIVITY OF THE NATURAL

Human Goods, Human Virtues, and Human Flourishing

Edited by

Mark J. Cherry

St. Edward's University
Austin, TX, USA

Springer

Editor
Mark J. Cherry
St. Edward's University
Department of Philosophy
Austin TX 78704
USA
markc@stedwards.edu

ISBN 978-90-481-2300-1 e-ISBN 978-90-481-2301-8

DOI 10.1007/978-90-481-2301-8

Library of Congress Control Number: 2009926879

© Springer Science+Business Media B.V. 2009
No part of this work may be reproduced, stored in a retrieval system, or transmitted
in any form or by any means, electronic, mechanical, photocopying, microfilming, recording
or otherwise, without written permission from the Publisher, with the exception
of any material supplied specifically for the purpose of being entered
and executed on a computer system, for exclusive use by the purchaser of the work.

Printed on acid-free paper

9 8 7 6 5 4 3 2 1

springer.com

Acknowledgements

The development of this volume benefited through the kind efforts of many friends and colleagues. Its origin was a series of lectures funded largely though a couple of generous grants from the Matchette Foundation, with additional funding from St. Edward's University. I am deeply thankful to the contributors, who recast their essays several times over the course of a couple of years, to craft the final versions contained herein. A special debt is owed to H. Tristram Engelhardt, Jr., Ana Iltis and Lisa Rasmussen, who are excellent colleagues and the best of friends.

I also wish to recognize the on-going generosity of St. Edward's University, the School of Humanities, and the Philosophy Department, especially Donna Jurick, SND, Louis T. Brusatti, William J. Zanardi, Peter Wake, Jack Green Musselman, and Stephen Dilley. Each has been instrumental, though in diverse capacities, to the success of this project. Thanks are also due to Gricelda Silva, who helped with the editorial process as my research assistant. As with all of my projects, this volume would not exist without the constant support, kindness, and love of my wife Mollie.

Contents

Contributors

Roberta M. Berry
School of Public Policy, Georgia Institute of Technology, Atlanta, GA, USA, robertaberry@gatech.edu

Mark J. Cherry
Department of Philosophy, St. Edward's University, Austin, TX, USA, markc@stedwards.edu

Stephen A. Erickson
Department of Philosophy, Pomona College, Claremont, CA, USA, Stephen_Erickson@pomona.edu

Anthony E. Giampietro, CSB
Department of Philosophy, University of St. Thomas, Houston, TX, USA, aegcsb@stthom.edu

Stephen S. Hanson
Department of Philosophy and Family and Geriatric Medicine, University of Louisville, Louisville, KY, USA, stephen.hanson@louisville.edu

Douglas V. Henry
Great Texts Program, Baylor University, Waco, TX, USA, Douglas_Henry@baylor.edu

Patrick Lee
Franciscan University of Steubenville, Steubenville, OH, USA, plee@franciscan.edu

Angela McKay
Department of Philosophy, The Catholic University of America, Washington, DC, USA, mckay@cua.edu

Jack Green Musselman
Department of Philosophy and Center for Ethics and Leadership, St. Edward's University, Austin, TX, USA, jackgm@stedwards.edu

Ian Nyberg
Department of Philosophy, University of Texas, Austin, TX, USA, inyberg@mail.utexas.edu

Christopher Tollefsen
Department of Philosophy, University of South Carolina, Columbia, SC, USA, Tollefsen@sc.edu

Peter Wake
Department of Philosophy, St. Edward's University, Austin, TX, USA, peteraw@stedwards.edu

William J. Zanardi
Department of Philosophy, St. Edward's University, Austin, TX, USA, williamz@stedwards.edu

The Normativity of the Natural: Can Philosophers Pull Morality Out of the Magic Hat of Human Nature?

Mark J. Cherry

1 Introduction

Western philosophy has long nurtured the hope definitively to resolve ethical and political controversies through appeal to right reason, and thereby to secure moral direction and human meaning without the need for a defining encounter with God or the transcendent. As an intellectual aspiration the expectation is for a moral rationality that is universal, secular, open to all, and able adequately to frame and guide the moral life. The roots for such optimism lie at the beginning of the second millennium, grounded in part in the Stoic, natural law, themes that characterized post-traditional periods in Hellenic culture and the late Roman Republic.[1] Appeals to discursive rationality became salient again in the second millennium as Western European thinkers framed what they presumed would provide a rational unity for their diversity of cultures, a unity not grounded in an encounter with God as Law-giver (e.g., God's giving seven commandments to Noah and his sons, the Gentiles, and 613 to Moses and the Jews) or dependent on theologians as mystics. Instead, moral and cultural unity was sought though rational philosophical reflection on human nature and the basic goods of a properly nurtured and virtuous human life—that is, through appeal to what has come to be called the natural law.[2]

Natural law theory holds that objective moral truths are not the product of cultural practice, social convention, individual choice, or religious conviction. Rather, the natural law seeks moral norms that are universally applicable—its content-full moral claims are validly addressed to anyone facing a choice upon which the principles and norms bear—and universally knowable—its canonical moral content is accessible to human beings generally (Boyle, 2004, p. 2). All persons are in principle capable of knowing and acting in accordance with its moral norms in virtue of their common capacity for reason. Natural law's moral epistemology assumes that persons are capable of appreciating, and should be held responsible for, the universal moral truths disclosed through practical reasoning regarding human nature

M.J. Cherry (✉)
Department of Philosophy, St. Edward's University, Austin,
TX 78704, USA
e-mail: markc@stedwards.edu

M.J. Cherry (ed.), *The Normativity of the Natural*, Philosophical Studies in Contemporary Culture 16, DOI 10.1007/978-90-481-2301-8_1,
© Springer Science+Business Media B.V. 2009

and basic human goods. This volume is a contribution to the growing literature of critical reflection on natural law theory.[3]

The natural law seeks to address permissible moral choice and social policy through an objective understanding of human nature and human good. It thereby seeks principles and norms for morality and law whose prescriptive force is not dependent for validity on human decision, democratic majoritarian decision making, past tradition, or social convention (Boyle, 1999). Thomas Aquinas, for example, held that the natural law is a function of reason "...promulgated by the very fact that God instilled it into man's mind so as to be known by him naturally" (*ST*, I-II, Q90, A4). The claim is that there is an objective good for human beings that can be articulated through reason itself, which in turn justifies the general canons of moral behavior. Accordingly, persons are obligated to will and to act in ways that are compatible with creating and integrating the basic human goods into their lives and the lives of others. Such goods thereby provide the basis for practical reasoning about virtuous choices as well as immediate reasons for action.

For example, in this volume Christopher Tollefsen and Patrick Lee each appreciate natural law moral norms as facilitating the attainment of the goal or telos of human life: full human flourishing. Similarly, Anthony Giampietro and Douglas Henry each argue that living a moral life regards the intentional shaping of one's character so that one instantiates and desires those goods central to human fulfillment. Here, the goal is not the maximization of such goods, but the making of rational choices in the pursuit of a virtuous, flourishing, human life. Not all choices are permissible, since many goals and actions, however instrumentally useful, are incompatible with basic human goods. For example, natural law theorists argue that intentionally killing the innocent is morally prohibited—it is intrinsically evil—since directly willing to kill the innocent is never compatible with human flourishing (Boyle, 2002). The positive norms compatible with basic human goods should be taught and habituated until they are embodied and lived, such that what is desired and what is morally required become identical.

Among the challenges natural law theorists face is the articulation and defense of a unique and canonical moral anthropology. Empirically, instead of unity, one finds considerable plurality of often contradictory and non-reducible religious and secular accounts of the goods central to human flourishing, as well as significantly diverse theories for rationally debating the merits of these divergent understandings of human nature. There appear to be at least as many competing moral anthropologies as there are major world religions and secular worldviews. Whose account of human nature and which view of human flourishing should be appreciated as morally normative? It may not be possible to read such judgments off of nature itself. For example, even if all agree that intrinsically evil acts violate and radically contradict the good of the person, all may not agree regarding which acts are, in fact, intrinsically evil. Articulating an answer requires an actual determination of facts as well as specification of the criteria and standards of rational choice employed. As Ian Nyberg argues, in this volume, securing a particular account of human nature and its core goods requires specification of moral content that will not be acknowledged as authoritative among those who do not already share a common moral anthropology.

Indeed, as Peter Wake and Stephen Erickson each argue, in their contributions, there might not even be a rationally defensible and morally definitive account of human nature. It is this tightly knotted set of questions to which this volume is addressed.

2 Thomistic Foundations: Natural Law Theory, Synderesis, and Practical Reason

Contemporary natural law theorists, such as Tollefsen and Lee, typically describe the categories of basic human goods as including health and safety, knowledge and appreciation of beauty, excellence in work and play, as well as friendship, peace and fraternity.[4] Such goods are appreciated as both universal to all human beings and as central to human flourishing, properly understood: "The basic goods, by contrast, are those that can be pursued for their own sake, and which just as such fulfill us in certain respects as human persons. Moreover, such goods perfect and fulfill all human beings" (Tollesfsen, p. 21). The work to be accomplished, as Tollefsen is aware, is not trivial: first, to defend the particular list of basic human goods; second, properly to articulate their content; and, third, to draw out their moral implications. Detailed arguments are necessary, for example, to connect basic human goods to the philosophical defense of natural law restrictions on the use of certain forms of biotechnology, such as reproductive cloning and genetic engineering.

Angela McKay argues, for example, that the natural law does not provide much practical moral guidance in itself—the natural law requires additional rationally defensible precepts and the personal inculcation of moral virtues to guide judgment. As she interprets Aquinas, "Important as the injunction to 'do good and avoid evil' might be, it gives us very little genuine guidance. Indeed, it is hard to see how even the further precepts Aquinas offers, namely that one ought to do what is conducive to the preservation of one's life and of the species, could provide one with much real guidance in practical affairs" (p. 43). The natural law, she argues, gives man his core orientation towards basic human goods. For genuine moral guidance one must cultivate the virtues—central personal characteristics necessary for human flourishing. The primary source of the moral virtues is "synderesis", a natural habitual knowledge regarding the first principles of action orienting man towards that which is good.

On this point, Lee argues that the first principles of practical reason and morality are self-evident; they are not derived from other propositions but rather arise out of insights into human nature. "One could put it this way: they are *ontologically grounded* in human nature even though they are not *logically derived* from propositions about human nature" (2008, p. 50). Knowledge of such goods is experiential; it occurs as the result of reflection on experiences that are genuinely fulfilling. We come to understand the good of health, for example, because of our experiences of being ill and healthy. Similarly, we experience the goods of friendship and peace, in contrast with hatred and violence. Each good is experienced as such and, over time, is habituated so that we come to understand it as a natural object worthy of pursuit:

> The natural inclinations are, in a way, data for the practical insight. How this occurs may be illustrated, I think, in the following way. As a young child I experience being healthy as opposed to being sick or having my knees scraped up or having burnt fingers. I *enjoy*, or *take delight in*, being healthy, and I *dislike*, or *have an aversion to* being sick or wounded. ... But at some point as a child I go further and I come to *understand* that being healthy is a condition worth pursuing, and being sick or wounded is a condition I should take steps to avoid. What has occurred here is a practical insight, an act of intellect, an insight that being healthy is fulfilling and thus worthy of pursuit, protection, deliberation about, and so forth (p. 51).

Such reflection, Lee argues, serve as the basis for practical deliberation about particular goods, thereby developing practical wisdom and guiding moral actions.

As part of his contribution, Tollefsen musters natural law arguments in defense of both the traditional marriage of man and woman, as well as natural procreation. Such goods, he argues, are not merely the reflection of a particular religious understanding but are, in fact, essential to human well-being. Marriage and reproduction are other-related goods, dependent on the interpersonal reality of human nature and human fulfillment. Tollefsen is critical of reproductive cloning and similar artificial reproductive biotechnology:

> To narrow our reproductive world in this self-focused way, by becoming autonomous procreators, seems reductive of our possibilities for well-being, by removing us from those others with whom we would have had the potential for interpersonal communion. Cloning is akin to masturbating for children, a lonely and solipsistic business; a world in which all reproduction had become disjoined from sex would be similarly depersonalized ... (2008, p. 23).

Outside of the traditional marital union of man and woman sexual activity becomes more or less recreational and children become the product of technical design. Tollefsen argues that permissible sexual activity is appropriately morally restricted to traditional monogamous heterosexual marriage. Children created through cloning, and other artificial reproductive techniques, fail to be the product of a loving, deeply human, relationship. In such circumstances, children are simply the outcome of a mechanical enterprise designed to manufacture progeny equipped with specific physical features, personal characteristics, and genetic options. Human reproduction appears much like shopping for an automobile, cell phone, or other accoutrement of modern living. Here, children are "...ultimately, in the eyes of their creators, artifacts, products of their will and design, rather than persons" (p. 23). He concludes that cloning, and any other biotechnical design and manufacture of children, should be rejected as contrary to basic human well-being.

Jack Green Musselman's challenge to such traditional moral conclusions is that natural law theorists, such as Tollesfen and Lee, must already presuppose a particular account of human flourishing, and the goods necessary for human well-being, to secure their particular judgments. There does not appear to be a single authoritative account of the characteristics of the appropriately flourishing human life, i.e., of basic human goods or of true integral human fulfillment, knowable through human reason. Western culture routinely confronts a fundamental fragmentation of moral visions regarding human goods and human flourishing. Contemporary western society, much like the ancient world, discloses a polytheism of moral visions; it is the seemingly irreducibility of this empirical reality that is frequently characterized as post-modernity. Alasdair MacIntyre, for example, describes western culture as no longer in possession of a common understanding of moral virtue (1981).[5] Contemporary moral discussions are embedded in what MacIntyre characterizes as a cosmopolitan culture articulated in the international language of secular modernity; that is, a morality isolated from traditional cultures and religions and devoid of a shared ethics to provide a common moral compass.

Consider, for example, Green Musselman's reflections on the life of the "heterosexual New York bachelor John", whose "non-marital sexual acts" he claims provide John with fulfilling intimacy goods.

> John may also thereby know he does not enjoy restricting his achievement of these goods with another person by abstinence. That is, even outside the bonds of marriage one might argue that our experience of such non-martial sexual acts helps us, after the fact when we use the intellect, to understand that we can realize these goods through direct experience (p. 64).

As Green Musselman's example illustrates, there has been a systematic calling into question, if not outright abandonment, of the traditional values and institutions that had previously grounded understandings of moral truth and human flourishing. It has become typical for individuals to cohabitate and to lead active sex lives outside of the traditional married life. As a result, children and adults are no longer educated, or nurtured, as virtuous beings: for example, to regard sexual activity as properly restricted to the monogamous marriage of man and woman. As a result, it is unclear why persons reflecting on natural habitual goods will necessarily come to the Roman Catholic moral judgments regarding marriage, sexual activity, and reproduction, which Tollefsen and Lee affirm. It is even less obvious that such judgments are "self-evident" given careful reflection on human nature and human experiences. Insofar as the background culture affirms the goods of sexual freedom and sexual pleasure among consensual adult partners, then "free love" and sexual experimentation will be the taken-for-granted practice upon which one will reflect. Very likely, there will cease to be any rationally meaningful moral context through which to differentiate appropriate from inappropriate sexual relationships aside from consent.

As Green Musselman makes clear, traditional Christian accounts of the truly flourishing human life, including appropriate uses of sexuality and the centrality of traditional marriage, have been marginalized. The result has been a transformation and fragmentation of the religious and cultural assumptions that once framed western moral judgments central to historical articulations of natural law moral theory. Such fragmentation, as Musselman illustrates, has set the stage for the emergence of deep-seated differences that divide morality and accounts of human flourishing into not merely different, but mutually antagonistic accounts of proper moral conduct.

3 Human Goods and Human Flourishing: Revitalizing a Fallen Moral Culture

Endeavoring to bridge the gap among diverse moral communities, governments have attempted to establish a universal moral culture by legal decree. There is a concern to legislate and regulate a single moral culture that speaks with authority to all. The establishment of moral values and goals is thus evermore being thrust into the political arena. For example, the very secular cultures of many Western countries seek legally, culturally and politically to marginalize the traditional moral assumptions that had once guided public life. Consider just a few examples: Spain, Belgium, and the Netherlands, among other nations, have legalized same gender marriage[6]; Belgium[7] and the Netherlands[8] among others have legalized physician assisted-suicide; abortion has become a more or less routine medical practice;[9] artificial reproductive technologies are sufficiently common that many techniques are thought of as part and parcel of the standard of medical care (Rae, 2003). Whereas Aquinas could draw on the richness of a background religious morality in ways that are simply unavailable to contemporary Western philosophers, Europe and North America cannot honestly claim that they represent Christendom: a Christian community with a single, content-full, coherent, and all-encompassing morality. Instead, each contains a plurality of religious and secular moral cultures. The result is a cardinal difficulty in providing a morally authoritative foundation for the content that is legally promulgated. As William Zanardi reflects on in his contribution, having diagnosed such deep moral shifts, how should one proceed to revitalize a culture gone astray? It is this question to which the second brace of essays is addressed.

Here Henry and Giampietro each turn to philosophy for personal consolation and cultural revitalization.[10] Henry, for example, seeks a rightly ordered and detailed philosophical anthropology to guide moral judgment: "... philosophy succeeds where therapeutic drivel cannot precisely when it discerns what we are within the context of our source and our end. By so doing, knowledge of what we are offers morally significant direction for our lives" (p. 78). Henry's specific targets are scientific rationalism, which seeks to reduce all that is real to only physical properties, and misologistic nihilism, the despair of finding any meaning to life through human reason. Scientific rationalists, such as Daniel Dennett, embrace a naturalistic metaphysics that reduces all of human nature to physics:

> A contemporary scientific rationalist's account of what we are might therefore begin broadly with sociology, but from that rough-grained picture of human nature would work down through psychology, biology, and molecular biology to chemistry, and from there would drill right on down to the way things "really are" as described by physics (p. 79).

On the other hand, misologistic nihilists, such as Albert Camus and Jean-Paul Sartre, both deny the existence of a human nature and, at the same time, despair of reason's ability to provide adequate guidance on human nature. Either stance, Henry argues, is incapable of adequately accounting for the depth of the human condition.

Similarly, Giampietro notes that traditional natural law theory appreciates man as possessing a telos that comes from God; it is not self created. He argues that philosophy is able to provide reasons for thinking that there is unity and purpose in the universe and "that there is a natural moral law that, if human beings follow, will point toward a lasting happiness. And philosophy can thus give good reasons for acting morally, even if one must suffer death in doing so" (p. 105). Giampietro concludes, however, that philosophy can only take us so far; it cannot provide the full cure necessary for the human condition. Reason in itself cannot provide deep and unique human meaning. Philosophy can help persons prepare for something more, but true healing must come from God, Himself.

Consider, as a practical example, adequate preparation for death. To state the matter bluntly: without an adequate theology to guide appropriate spiritual preparation for the end of one's life, death is simply banal. For the devoutly secular, the preferred death occurs by choice or without warning; in either case, perhaps after good financial planning, but without the labors of spiritual preparation. Death is simply the end; the end of all experience, the end of all meaning. In contrast, within an authentic Christian appreciation of medicine, life-and-death decision making is integrated with Christian asceticism. Traditional Christians pray for a foreseen death with adequate time for confession and repentance. As Christ's parable about the foolish rich man makes clear: careful, spiritual, preparation for death is essential. At the time of his death, the foolish rich man hears a voice from God saying " 'You fool, this night your life will be demanded of you; and the things you have prepared, to whom will they belong?' Thus will it be for the one who stores up treasure for himself but is not rich in what matters to God" (Luke, 12: 20–21). Christians know that the soul must be adequately prepared. Such preparation cannot come from discursive reason or rationalistic philosophy. But here, Henry will likely agree. As he concludes, "When all is said and done, then, the question of whether or not human nature is morally normative, and in what ways, may depend entirely upon whether one does or does not find it possible to embrace a faith beyond reason" (p. 94).

4 The Malleability of Human Nature and the Morality of the Natural

As Henry and Giampietro each note, one of the core challenges to natural law theory is the claim that there is not a "human nature" as such, which would also imply that there is not a universal understanding of full human flourishing. Stephen Erickson, Peter Wake, and Roberta Berry each pursue aspects of this line of argument. Erickson argues, for example, that human nature itself is malleable. He notes that fundamental aspects of human nature, such as brain chemistry, are changeable. From a scientific perspective, the human brain is the product of evolution; moreover, the brain continues to evolve in response to current environmental stimuli. Such shifts over time will impact our conceptualization of human nature; it will likewise change our appreciation of human nature and the goods appropriate to a fulfilling human life. Here he raises an essential question: "The expansion of our cerebral cortex over many millennia, for example, has been extraordinary. Is there a particularly strong reason to remain consciously passive in such matters just at the time that various interventions become possible that can accelerate and direct further evolutionary change?" (Erickson, p. 134). To rephrase Erickson's question: Why should we passively accept changes to human nature caused by potentially random environmental stimuli, when we could actively direct human evolution, utilizing the tools of both imagination and biotechnology? As he notes, some individuals may have interest in "upgrading" themselves—e.g., utilizing nanotechnology for human computer interfaces or genetic engineering—and some will possess adequate private resources to sponsor such experiments on themselves or with consenting others. Are there sufficiently compelling reasons to forego or coercively to prevent such willing human experimentation?

Wake both echoes and further develops key threads of Erickson's argument throughout his crisp articulation that human nature is not a given. Human nature, he contends, should be thought of in terms of "plasticity": "there is nothing that is already there, for self-actualization is not a goal that can be predetermined" (p. 148). As free beings, persons determine their nature and their self-actualization, by way of free choices and free actions. Humans have what he terms, drawing on Hegel, the power of the "pure I" to negate what is (p. 139). This is a power to reflect on ourselves as well as to distance ourselves from pre-conceived notions of human nature, to shape ourselves through our free choices of who and what we wish to become. As he puts it: "Practically, it means we are a work in progress, we are underway, we are free" (p. 140). Much as humans shape their environments to suit their particular interests, humans also change their character and nature, to suit interests and goals. It is this ability to question ourselves and shape ourselves—a movement between nature and artifice—that is proper to humans; as a result, the basic goods central to human life and to human flourishing cannot be defined or determined in advance of the free choices of persons.

The challenges for natural law theory are significant. Even if it is possible to determine what is "natural" to humans or to detail an underlying "given" of human nature, as Nyberg argues, by itself this would be insufficient to ground particular moral judgments. Why should nature, even human nature, be thought of as a moral boundary beyond which one must not trespass? Perhaps nature is a challenge to be addressed, overcome, and set aside. Nyberg argues that even assuming that what is natural is somehow "best" is a misunderstanding of biological and evolutionary processes:

> Observing humans reveals all sorts of behaviors, but we still want to call some of them good and some of them bad. Indeed, much of ethics is precisely about constraining natural behaviors. Murder, rape, adultery, and theft, for example, are so ubiquitous in human society that many scientists have attempted to give natural selection accounts of these phenomena (p. 184).

There may be good grounds for believing that such actions possess advantages for natural selection, such as increasing reproductive fitness (e.g., rape to create more progeny; murder of reproductive rivals, and so forth); however, such advantages would not thereby morally justify the behavior. Often morality requires one to set aside and control one's "natural" passions, so as to act with a more virtuous character. Nyberg critiques five different definitions of "unnatural": (i) "what is not practiced by other animals"; (ii) "what is disgusting or offensive"; (iii) "what is unusual or abnormal"; (iv) "what violates an organ's principal purpose"; and (v) "what does not proceed from innate desire", showing in each case that an account of human nature is ambiguous with regards to particular moral judgments.

In short, even if man is able adequately to discover and specify the "laws of nature", such "laws" in themselves would not be morally normative. The laws of nature just are (see Engelhardt, 1996, p. 56). As Berry notes throughout her exploration of the potential for a genetically engineered post-human future and its implications for how those beings might rethink moral virtues, human nature, and its implications for morality, is remarkably opaque. One must possess an independent set of moral standards to judge which of the general tendencies of nature (1) ought to be acknowledged as appropriate limits on human behavior, (2) are good or virtuous and thus to be morally affirmed, (3) are sinful or vicious and thus to be morally denounced, and (4) ought to be overcome or set aside through human effort and ingenuity.

5 Conclusion

The dominant strategy for addressing ethical and political controversies, such as marriage, genetic engineering, or reproductive cloning, has been the search for secular philosophical standards of moral reasoning or universal guidelines, such as "human rights" or "human dignity", so as to ground a fully secular moral anthropology or to claim the existence of a "moral consensus" and thereby to foreclose debate, while also asserting the existence of authoritative moral content to guide individual choice and policy formation. "Human dignity" and "human rights", for example, are frequently cited as possessing universal normative relevance as a standard against which to assess competing moral claims, and thereby to resolve moral controversies. Consider:

> Perhaps the two most distinctive features of international instruments relating to biomedicine are the very central role given to the notion of "human dignity" and the integration of the common standards that are adopted into a human rights framework. This is not surprising if we consider that human dignity is one of the few common values in our world of philosophical pluralism. Moreover, in our time, a widespread assumption is that the "inherent dignity . . . of all members of the human family" is the ground of human rights and democracy [United Nations, *Universal Declaration on Human Rights, Preamble*]. It is indeed difficult, if not impossible, to provide a justification of human rights without making some reference, at least implicitly, to the idea of human dignity (Andorno, 2002, p. 960).

Here "human dignity" is being cited as a universal value, which while vague, nonetheless provides justification for substantive far-reaching social/political obligations. Through its appeal to purportedly shared core values, e.g., "human dignity", and to supposedly universally valid core principles, e.g., "human rights" and "social justice", the very secular field of philosophical ethics claims to have transcended cultural and religious differences, binding all peoples and nations. The expectation is that through such universal values, secular moral rationality discloses a communality of all persons, justified neither in faith nor in divine revelation, but through secular reason. General philosophical reason is presumed to provide access to a moral foundation for human community, justifiable without appeal to particular religions or cultures.

The challenge for natural law theory, as Green Musselman, Berry, Erickson, and Wake each illustrate, is that general secular reason is unlikely to support the traditional moral positions of natural law theory. Western secular philosophical ethics continues to give political and institutional expression to the Enlightenment expectation that ethics, and thus bioethics, should liberate individuals from those customs and constraints held to be contrary to the demands of universal secular moral reason. Choosing other than in accord with general secular reason, in adherence to traditional religious commitments, for example, is characterized as acting under a false consciousness or as the victim of exploitation, social, cultural or patriarchal despotism. Proponents do not characterize such harmonization across diverse segments of national and international populations as limiting legitimate diversity or restricting individual liberty but rather as returning individuals, families, and communities, to the appropriately objective standards of rationally disclosed morality. In this fashion, "ethicists" typically function as "conceptive ideologists" elaborating a national and international ethics in support of a liberal secular political ideology (Engelhardt et al., 2006), to guide court decisions, public deliberation, clinical decision-making, and legislative action, as well as international convention and treaty. Such a liberal secular political ideology is routinely hostile to the moral norms typical of traditional natural law theory.

This contemporary hunger for "global moral consensus", even in the face of apparently irresolvable moral pluralism, expresses a deep seated need to deny the existence of real moral difference, binding all humans together in a common morality, through the social and political mechanism of announcing the secular equivalent of orthodox religious belief. Recasting moral and political controversies, such as appropriate expressions of human sexuality, within the rhetorical arena of "human rights," "human dignity," and a purported "moral consensus" offers a rhetorically powerful language as a political tool to forestall and foreclose debate and to impose particular secular moral viewpoints. Significant moral disagreement is to be shunned, if not actively persecuted.

Which (or whose) metaphysical assumptions regarding human nature should guide the foundations of contemporary culture and public policy? Which (or whose) should ground moral judgments? Should marriage be seen as the union of one man and one woman or should the state endorse attempts at same gender contractual marriages? Should sexual activity be understood as properly restricted to the marriage of one man and one woman or should modern culture embrace free-love and recreational sexual experimentation among consenting adults? Should human reproduction be appreciated as appropriately the result of the carnal union of husband and wife or should medicine embrace reproductive cloning? Each of these debates, among many others, is a battleground in the on-going culture wars.[11] Framing public moral culture and political policy is never neutral. This volume seeks to clarify and critically explore such deep and thorny philosophical issues.

Notes

1. Consider: "*Atticus.* —In your opinion, then, it is not in the edict of the magistrate, as the majority of our modern lawyers pretend, nor in the rules of the Twelve Tables of our Statutes, as the ancient Romans maintained, but in the sublimest doctrines of philosophy, we must seek the true source and obligation of jurisprudence.

   Marcus.— ... But the subject of our present discussion soars far higher, and comprehends the universal principles of equity and law. In such a discussion therefore on the great moral law of nature, the practice of the civil law can occupy but an insignificant and subordinate station. For according to our idea, we shall have to explain the true nature of moral justice, which is congenial and correspondent with the true nature of man. We shall have to examine those principles of legislation by which all political states should be governed ..." (Cicero, *Treatise On the Laws*, I, 18). Cicero (B.C. 106–143) develops an account of natural law as right reason.
2. Consider, for example, Thomas Aquinas (c. A.D. 1227–1274) *ST* II, 2, Q 94, or the works of Francisco de Victoria (c. A.D. 1480–1546). See also Midgley's *The Natural Law Tradition and the Theory of International Relations* (1976); Hittinger's *The First Grace* (2003); Cherry's *Natural Law and the Possibility of a Global Ethics* (2004) and *The Death of Metaphysics; The Death of Culture* (2006).
3. As H.L.A. Hart rightly noted, natural law theory reaches beyond religious discussions to secular philosophy: "Natural Law has ... not always been associated with belief in a Divine Governor or Lawgiver of the universe, and even where it has been, its characteristic tenets have not been logically dependent on that belief. Both the relevant sense of the word 'natural,' which enters into Natural Law, and its general outlook minimizing the difference ... between prescriptive and descriptive laws, have their roots in Greek thought which was, for this purpose, quite secular. Indeed, the continued reassertion of some form of Natural Law doctrine is due in part to the fact that its appeal is independent of both divine and human authority ..." (1961, p. 183f).
4. Boyle and his collaborators describe the categories of basic goods as follows: "As *animate*, human persons are living organic substances. Life itself—its maintenance and transmission—health, and safety are one form of basic human good .... As *rational*, human beings can know reality and appreciate beauty and whatever intensely engages their capacities to know and feel, and to integrate the two. Knowledge and aesthetic experience are another category of basic good. As simultaneously *rational* and *animal*, human persons can transform the natural world by using realities, beginning with their own bodily selves, to express meanings and/or serve purposes within human cultures. Such bestowing of meaning and value can be realized in diverse degrees; its fullness is another category of basic good: excellence in work or play. ... As *agents through deliberation and choice*, they can strive to avoid or overcome various forms of conflict and alienation, and can seek after various forms of harmony, integration, and community (fellowship). ... Most obvious among the basic human goods of this relational dimension are various forms of harmony between and among persons and groups of persons: friendship, peace, fraternity, and so on" (1987, pp. 279–280).
5. "The social and cultural condition of those who speak that kind of language [is] a certain type of rootless cosmopolitanism, the condition of those who aspiring to be at home anywhere—except that is, of course, in what they regard as the backward, outmoded, undeveloped cultures of traditions—are therefore in an important way citizens of nowhere... It is the fate toward which modernity moves precisely insofar as it successfully modernizes itself and others by emancipating itself from social, cultural, and linguistic particularity and so from tradition" (MacIntyre, 1988, p. 388).
6. The Netherlands: see Staatsblad 2001 No. 9, Art 1.30(1) of the Dutch civil code which reads: 'a marriage can be entered into by two persons of different or same sex'; Belgium: see Moniteur Belge (Official Gazette) 28 Feb 2003 ed. 3, pp. 9880–9883; Spain: see Law 13/2005, of 1 July, providing for the amendment of the Civil Code with regard to the right to contract marriage, Boletin Oficial del Estado no. 157, 2 July 2005, pp. 23632–23634, http://www.boe.es/boe/dias/2005-07-02/pdfs/A23632-23634.pdf (in force 3 July 2005): Article 44; see generally Fiorini, Aude, "New Belgian Law on Same Sex Marriage and the PIL Implications," *International & Comparative Law Quarterly*, 52, 1039–1049 (2003).

7. See Jean-Louis Vincent, "End-of-life practice in Belgium and the new euthanasia law," *Intensive Care Medicine* 32(11), 1908–1911 (2006).
8. Geritt K. Kimsma and Evert van Leeuwen, "The role of family in euthanasia decision making," *HEC Forum* 19(4), 362–370 (2007).
9. According to some studies the numbers range from approximately just under 900,000 (Centers for Disease Control) to 1.3 million (Alan Guttmacher Institute) abortions a year in the U.S. See Strauss et al. 2003; www.nrlc.org/abortion/facts/abortionstats.html
10. Consider John Paul II's reflections on a similar question in *Fides et Ratio*: "*Moral theology* has perhaps an even greater need of philosophy's contribution. In the New Testament, human life is much less governed by prescriptions than in the Old Testament. Life in the Spirit leads believers to a freedom and responsibility which surpass the Law. Yet the Gospel and the Apostolic writings still set forth both general principles of Christian conduct and specific teachings and precepts. In order to apply these to the particular circumstances of individual and communal life, Christians must be able fully to engage their conscience and the power of their reason. In other words, moral theology requires a sound philosophical vision of human nature and society, as well as of the general principles of ethical decision-making" (1998, p. 100).
11. James Davidson Hunter coined the term "culture wars" in his 1991 study *Culture Wars: The Struggle to Define America*.

Bibliography

Andorno, R. (2002). Biomedicine and international human rights law: In search of a global consensus. *Bulletin of the World Health Organization*, 80(12), 959–963.

Aquinas, T. (1947). *Summa theologiae*. New York: Benziger Brothers.

Berry, R. (2009). The posthumanist challenge to a partly naturalized virtue ethics, in M. J. Cherry (ed.), *The normativity of the natural: human goods, human virtues, and human flourishing*. Dordrecht: Springer.

Boyle, J. (1999). Personal responsibility and freedom in health care: A contemporary natural law perspective, in M. J. Cherry (ed.), *Persons and their bodies: rights, responsibilities, relationships* (pp. 111–142). Dordrecht: Kluwer Academic Publishers.

Boyle, J. (2002). Limiting access to health care: A traditional Roman Catholic analysis, in H. T. Engelhardt, Jr. and M. J. Cherry (eds.), *Allocating scarce medical resources: roman catholic perspectives* (pp. 77–95). Washington, DC: Georgetown University Press.

Boyle, J. (2004). Natural law and global ethics, in M. J. Cherry (ed.), *Natural law and the possibility of a global ethics* (pp. 1–15). Dordrecht: Kluwer Academic Publishers.

Engelhardt, H.T., Jr., (1996). *The foundations of bioethics*, second edition. New York: Oxford University Press.

Engelhardt, H.T., Jr., Garrett, J. and Jotterand, F. (2006). Bioethics and the philosophy of medicine: A thirty-year perspective. *The journal of medicine and philosophy*, 31(6), 565–568.

Erickson, S. (2009). Reflections on secular foundationalism and our human future, in M. J. Cherry (ed.), *The normativity of the natural: human goods, human virtues, and human flourishing*. Dordrecht: Springer.

Finnis, J., Boyle, J. and Grisez, G. (1987). *Nuclear deterrence, morality, and realism*. Oxford: Clarendon Press.

Cherry, M. J. (2004). *Natural law and the possibility of a global ethics*. Dordrecht: Kluwer Academic Publishers.

Cherry, M. J. (2006). *The death of metaphysics; the death of culture*. Dordrecht: Springer.

Cicero, M.T. (1841–1842). *The political works of marcus tullius cicero: comprising his treatise on the commonwealth; and his treatise on the laws*, F. Barham, Esq. (trans.). London: Edmund Spettigue. [On-line.] Available: http://oll.libertyfund.org/Home3/Book.php?recordID=0044.02

Giampietro, A. (2009). Preparation for the cure, in M. J. Cherry (ed.), *The normativity of the natural: human goods, human virtues, and human flourishing*. Dordrecht: Springer.

Green Musselman, J. (2009). Natural law for teaching ethics: An essential tool and not a seamless web, in M. J. Cherry (ed.), *The normativity of the natural: human goods, human virtues, and human flourishing*. Dordrecht: Springer.

Hanson, S. (2009). Moral acquaintances and natural facts in the Darwinian age, in M. J. Cherry (ed.), *The normativity of the natural: human goods, human virtues, and human flourishing*. Dordrecht: Springer.

Hart, H. L. A. (1961). *The concept of law*. Oxford: Oxford University Press.

Henry, D. (2009). Quid Ipse Sis Nosse Desisti, in M. J. Cherry (ed.), *The normativity of the natural: human goods, human virtues, and human flourishing*. Dordrecht: Springer.

Hittinger, R. (2003). *The first grace: rediscovering the natural law in a post-christian world*. Wilmington: ISI Books.

Hunter, J. D. (1991). *Culture wars: the struggle to define america*. New York: Basic Books.

Lee, P. (2009). Human nature and moral goodness, in M. J. Cherry (ed.), *The normativity of the natural: human goods, human virtues, and human flourishing*. Dordrecht: Springer.

MacIntyre, A. (1981). *After virtue*. Notre Dame: University of Notre Dame Press.

MacIntyre, A. (1988). *Whose justice? which rationality?* Notre Dame: University of Notre Dame Press.

McKay, A. (2009). Synderesis, law, and virtue, in M. J. Cherry (ed.), *The normativity of the natural: human goods, human virtues, and human flourishing*. Dordrecht: Springer.

Midgley, E. B. F. (1976). *The natural law tradition and the theory of international relations*. New York: Harper and Row.

Nyberg, I. (2009). Can moral norms be derived from nature? The incompatibility of natural scientific investigation and moral norm generation, in M. J. Cherry (ed.), *The normativity of the natural: human goods, human virtues, and human flourishing*. Dordrecht: Springer.

Paul, J. P., II (1998). *Fides et ratio*. Vatican City: Libreria Editrice Vaticana.

Rae, S. B. (2003). United States perspectives on assisted reproductive technologies, in J. F. Peppin and M. J. Cherry (eds.), *Regional perspectives in bioethics* (pp. 21–38). Lisse: Swets and Zeitlinger.

Strauss, L. T. et al. (2003). Abortion surveillance – United States, 2003, *Centers for disease control*, [On-line.] Available: http://www.cdc.gov/mmwr/preview/mmwrhtml/ss5511a1.htm

Tollefsen, C. (2009). Human nature and its limits, in M. J. Cherry (ed.), *The normativity of the natural: human goods, human virtues, and human flourishing*. Dordrecht: Springer.

Wake, P. (2009). Nature as second nature: Plasticity and habit, in M. J. Cherry (ed.), *The normativity of the natural: human goods, human virtues, and human flourishing*. Dordrecht: Springer.

Part I
Thomistic Foundations: Natural Law Theory, Synderesis and Practical Reason

Human Nature and Its Limits

Christopher Tollefsen

This paper is concerned with the prospect of changes in human nature. Various types of research—in genetics, nanotechnology, artificial intelligence, and robotics (GNR),[1] for example—are currently underway to provide enhancements to that nature. But many believe that some such changes, at least, would do more than merely enhance. Francis Fukuyama, for example, a member of the President's Council on Bioethics, has titled a recent book *Our Posthuman Future* (2002) indicating his worry that the fruit of such research will be a change in our very humanity; this worry is shared by his fellow Council member Charles Krauthammer (2001, p. 60). While any such metamorphosis must be considered a future contingent at best, present day research on such views brings us closer to the possibility minute by minute.

Defenders of research into these new technologies can respond to the Fukuyama-Krauthammer sorts of concerns about changes in human nature in two ways: one more and one less radical. More radically, they can embrace the possibilities of significant changes in human nature, viewing it as one more step of Darwinian evolution, perhaps, though in this case, a step deliberately brought about (see Silver, 1998). In its most radical embodiment, this proposal is part of the philosophic, social, and political agenda of the so-called "transhumanist" movement (Ramez, 2004; see also the various statements and documents available at www.transhumansim.org).[2]

This response is unlikely to placate objectors, but it should be addressed on its own terms: is it genuinely worrisome and morally problematic to contemplate significant changes to human nature? Could we not improve on human nature as we know it, and if so, would we not have an obligation to pursue such a possibility? Moreover, there are possible goals that fall short of radical change in nature, but that would involve some change that would seem inevitably to be for the better: if we could eliminate genetic tendencies to crime, or to alcoholism, for example, we would not have changed human nature as such, but we would certainly have radically altered its parameters, and some would argue, for the better.[3]

C. Tollefsen (✉)
Department of Philosophy, University of South Carolina, Columbia, SC, USA
e-mail: Tollefsen@sc.edu

M.J. Cherry (ed.), *The Normativity of the Natural*, Philosophical Studies in Contemporary Culture 16, DOI 10.1007/978-90-481-2301-8_2,

© Springer Science+Business Media B.V. 2009

This last question raises the second avenue of response, less radical than the wholehearted embracing of the opportunity to change human nature. There is a relatively murky line between enhancing our nature and changing it. Suppose, for example, that it becomes possible to download information into mechanically enhanced brains so as to lead to increased memory, or superior intelligence, or even simply virtual experiences. Would such a circumstance constitute a change in human nature?[4] If so, how would it differ from the use of prosthetic limbs, or even contact lenses? So even if we take our nature as a moral given, defenders of the new technologies could say, this still leaves a considerable amount of room for experimentation, ultimately with a view to introducing significant, but not nature-altering, changes in our identity.

Navigating this dispute requires an answer to the following question: Just what is our human nature, and what is the normative importance of that nature? With answers to these questions in hand, it will be possible to say something about what sorts of changes would constitute changes in human nature and what sorts of changes would be normatively inappropriate given our human nature.

To answer these questions, I propose to do the following. First, I will briefly map out two features of our human nature, which I will call, respectively, our animal identity and our practical identity. Second, I will identify three broad types of changes that could compromise our identity on either or both these axes of our nature, and one general type of change that would not compromise that identity. The third thing I will do is to suggest that at least some of the proposed changes would require experimentation and research that would, given the earlier identified features of our nature, be immoral, even if the proposed changes themselves would not be. So even where change in our nature, or mere enhancement of our nature, would itself be morally permissible, it will in many cases be impermissible to discover how best to make such changes.

1 Our Animal Nature

What sort of being are we? That is, what sort of being are you, the reader, and I, the author of this paper, and all other beings essentially like us? What sort of entity are we, such that if any of us ceased to be *that* sort of entity, we would cease to exist altogether? It is clear that none of us cease to exist when we enter or leave a room, or when we grow a year older; what we are is not, essentially, a function of our place or age. But what is our identity and our continued existence a function of?

Two answers are natural to this question. It is natural, first, to think that we are human beings, animals of the species *homo sapiens*. This answer is not simply a bit of metaphysical speculation, but is true to how we experience ourselves and our relations to other persons and to the world. For we experience ourselves, at a very commonsense level, as being bodily beings.[5]

Thus, when we run or walk, we do not think of our body as bringing us someplace, but of ourselves as going someplace. When we meet a new person, we do not think of the exchange as an interface of two bodies, mediating the communication of two *other* sorts of entity who cannot touch or directly communicate with each other. Rather, in seeing the other's face we think we have seen the other himself or herself. We think this of ourselves as well: when we look in the mirror, we believe that we see a reflection of ourselves.

The second natural answer is to say that we are persons. But this answer is ambiguous. Does "person" mark out a kind of entity different from the class of human beings? It is not uncommon for philosophers to answer affirmatively, and to assert that *we* are such entities, persons, not identical to material organisms. Such views allow our existence to be independent of the existence of the human being in various ways; the human being can come to exist before the human person, and may persist after the human person's demise. Perhaps the human person can exist independently of the human being, as a disembodied soul. All such views are dualist; they identify two entities in the vicinity of our lives, the bodily organism, and the non-bodily person, and assert our identity with the second, but not the first.

Dualism inevitably separates what seem to be integrated aspects of our lives into alienated parts. A dualist such as Plato or Descartes, for example, separates the mind or soul from the body, and thus distinguishes the inner self from the outer body. With such a separation, knowledge of the external world becomes very troublesome. If I am a mind, then the impact of the external world on my body in, for example, sight, is not an impact made by the world on me. So the effect made by the world upon the body must then in turn be made upon the mind by the body.

Now it is mysterious how substances of such different natures as minds and bodies can have causal impacts upon one another; but it is equally problematic to assume that whatever mental representation I have of the world as a result of the body's causation really is accurate to the world. For the Cartesian dualist, the mind is at such a distance from the world that knowledge begins to seem impossible. Thus many philosophers have believed that an immediate consequence of Descartes' dualism is the problem of knowledge, and the inevitability of skepticism.

The connection is similarly severed as regards the nature of human action. When I wish to eat an apple I reach out and take it; I then take a bite. Thus, I see, reach for, touch, and taste the apple. In all these actions, consciousness—mind—and body are intimately related. My seeing is not like the inner presentation of a picture. My reaching out does not consist of an inner attempt and then an external reach. Nor do touch and taste consist of an external sensation and then an internal one. Internal and external are closely integrated in all these happenings.[6]

For these and other reasons, it seems much more plausible to hold that we are human beings, physical organisms whose bodily existence is directly in contact with the rest of the physical world. But does this do violence to our natural understanding of ourselves as persons?

It does not. A person is a being of a sort that possesses the capacity for reason and choice. Persons can respond to reasons, as opposed to the mere pull or push of instinct and desire, by evaluating those reasons, and choosing in light of them. I will return to this point in the next section; but for now it is important to point out that these capacities just are part of our nature as human beings, as organisms of this particular sort.

This insight is easily lost sight of if we view a human organism by way of a snapshot, a picture of that organism at one particular time, say, the time prior to birth when the human being seems incapable of rational thought and deliberation. How can that sort of being be a person? But all organisms lead temporal lives, and the nature of those lives can only be understood by looking, not at a snapshot, but at the characteristic way in which such lives unfold in time. Dogs are not, from their beginnings, capable of eating meat, but they are by their nature carnivores. This truth about what dogs are is appreciated only by looking at the way dog lives unfold temporally.

Similarly, human beings are persons; they have a rational life that includes a sensitivity to reasons and an ability to respond rationally to reasons. This aspect of their nature, like the carnivorous nature of the dog, can only be appreciated by looking at the life of the human organism through time, and not merely at a time. That life is a personal life and, thus, we can say that human beings are persons, not because the category "person" identifies their true nature, but because that nature is truly personal (Tollefsen, 2006).

2 Our Practical Identity

It follows from the analysis of the previous section that the being of a human person is not given all at once; human persons are not everything that they are at any one time. But it also follows from the personal nature of human lives that what a human being is at a time is in part a function of what they can be. Because the nature of human beings involves deliberation and choice, human persons are in part self-constituting; what a particular human being is at a time is a function of what he or she has chosen to be. But this means that what human beings are as such is in part a function of what they *can* choose to be; human nature includes not just what is given in human existence, but also includes the horizon of reasons that orients what any individual human being can become.

What are the reasons for action, these horizons of self-constitution, for human beings? Following other theorists in the natural law tradition, I hold that the foundations of human action and self-constitution are to be found in the well-being and fulfillment of human persons and the communities they form. Moreover, natural law theory understands human fulfillment—the human good—as variegated. There are many irreducible dimensions of human well-being and flourishing; there are thus a variety of reasons that establish the more than given essence of human nature. I shall call such reasons that are based in human well-being "human goods."[7]

This is not to deny that human nature is determinate; rather, it is to affirm that our nature, though determinate, is complex. We are animals, but are also rational. We are individuals, but are also social. We both know and transform reality. The multiplicity of human goods which constitute the basic aspects of human well-being reflects this complexity of our nature: we could not be fulfilled if we are attentive to but one dimension of our well-being.

I shall call the most fundamental dimensions of our well-being *basic goods*. They are goods, obviously, because they are good for us, fulfill us, lead us to human flourishing. But they are basic because they give reasons for us to act that do not need grounding in some further reason. In this respect, basic goods are unlike other goods that are pursued only because they help us to attain or achieve something else. No one takes medicine just for its own sake, or pursues money just to possess it. Money and medicine are instrumental goods; they are pursued for the sake of something else.

The basic goods, by contrast, are those that can be pursued for their own sake, and which just as such fulfill us in certain respects as human persons. Moreover, such goods perfect and fulfill all human beings. So it is possible to act for the sake of a human good not just for one's own sake, but for the sake of others; and one has good reason to do so, as we shall see, because friendship and concern for others is itself a basic good, an aspect of human well-being.

The basic goods include the following. As animate, bodily creatures, our lives and our health are basic goods: we can act simply to promote or preserve life in our own person or that of others.

As rational, knowledge is, for us, a good to be pursued for its own sake; human beings are better off knowing, than in ignorance. Of course, knowledge, like all the basic goods, can also be pursued for the sake of something else; but, like life and health, the pursuit of knowledge for its own sake also gives meaning to our actions, a meaning that does not require some further justification. Similarly, human beings can take pleasure in works of art, and pursue aesthetic experience for its own sake.

Human beings not only can know, but can transform their world through action. They do this in work, and in play, which, when pursued as forms of excellence of performance, are humanly fulfilling. People can, again, work for the sake of something else, but no one needs a further reason to do good work or to play well than the goodness of work and play themselves.

Human beings are fulfilled not only alone, but in community. The goods of community include friendship, marriage between a man and a woman, and just and peaceful societies. Each of these goods contributes to human well-being in a way different from any other.

Two further goods complete the list. Each is similar to the goods of community in a way. All those goods require forms of harmony, for the wills of the members of the community, whether that community is a community of two friends, or of husband and wife, or of political equals, must be harmonized with one another. One's will must also be harmonized, on the one hand, with whatever source or sources of transcendent meaning exist in the universe and, on the other, with those aspects of one's self that can conflict with one's choices: the deliverances of practical reason, and one's emotions and dispositions. The good of religion corresponds to the former sort of harmony, the good of all around practical reasonableness or authenticity to the latter.

Our human nature thus emerges as having two crucial aspects. First, our nature is in part a practical nature: that is, it is a nature specified by our practical possibilities, in particular, by those basic goods that constitute the horizon of our possible flourishing. This practical nature itself has a dual aspect. On the one hand, our good, though of vast depth and breadth, is not infinitely plastic. That is, although any particular human good—life, knowledge, play, for example—each offers a vast number of ways of pursuit, promotion, and participation, nevertheless, the core list of the basic goods is finite, relatively small, and permanent. There are many ways of pursuing the good of life: doctors, lifeguards, law enforcement officials, the drafters of the Clean Air Act, all pursue this good, and even among these, there are, for example, many ways of being a doctor. But the status of life as a basic human good is inflexibly essential to our nature; and the list of basic goods seems fixed; there are no new goods "out there" waiting to be discovered.

The other aspect of our practical nature, beyond the essentially fixed horizon of possibilities, is that these possibilities must be brought about for us by our own action. It is true that some of these goods are, as Grisez puts it, substantive, rather than reflexive; the former do not require an agent's choice genuinely to benefit the agent. We are simply better off being alive, whether we have deliberately acted to sustain or promote our life, or not. Nonetheless, our *given* nature is largely one of capacities, which require our action to be brought to actuality. Our life must be a life of deliberation, choice, commitment and action if it is to be a good and flourishing life. Robert Nozick's experience machine thought experiment, to which I shall return shortly, helps us to appreciate this: we do not want our lives to be lives of merely passive benefit, of induced experiences, but lives of action, lives of which we are agents and authors. This realization lies behind the deep value we accord to autonomy and our strong requirement of informed consent in medicine and research.

Thus, the first two-sided point: our nature is practical, involving a necessary relation to a finite set of goods, and requiring action on our part to participate in those goods and flourish as selves. The second point, defended in the previous section, is that our nature is also an animal nature: we are not human spirits, essentially disembodied, but are human animals, notable for being spiritual animals in a way that other mammals are not, but animals, and animals of a particular sort, nonetheless. This aspect of our being is of course tied up with our practical nature in a wide variety of ways: it is our bodily life that is a good in itself, not simply the life of our soul. It is through our senses that we come to know, through our physical communion with others that we play, marry, raise children, and so on. Art is experienced through our bodily senses; integrity is a matter not just of doing, thinking, or willing one thing, but also of feeling in accordance with the dictates of reason. In short, our animal nature pervades our practical nature and vice versa. Any attempt to understand what we are that ignores or downgrades on or the other of these aspects, fails to capture the full reality of human nature.

3 Human Nature and Enhancement

There seem to me three ways in which enhancement and other technologies can threaten us in respect of our nature as described in the two previous sections: they may threaten the relationship between us and the finite set of goods that offer our fulfillment as human persons; they might threaten our ability practically to pursue those goods, and to participate in those goods through our own creative activity; and they might show disregard for the particularity of our animal nature, a disregard which would likely have consequences in regard to our relationship to the goods as well.

Consider, for example, research aimed at making humans autonomously, i.e., asexually, reproductive. Current technologies, such as in vitro fertilization and, to a greater extent, reproductive cloning, already have brought us part of the way towards such a state of affairs. The aim of such research is to remove from the sphere of sex, marriage, and family the act of creation of children and, indeed, to remove it from any necessarily interpersonal context at all (see Silver, 1998).[8] Forward looking thinkers such as Ray Kurzweil predict that within one hundred years, our species will no longer reproduce sexually at all; sex will become recreational and, indeed, mostly virtual. Procreation, meanwhile, will be essentially medical/technical.

Now, although such a situation might be taken by some to be a good thing, it is a different and a lesser thing than traditional procreation in the context of a marriage. It is precisely here, of course, that arguments become necessary but difficult. Two considerations seem important. First, sex and marriage are other-related, interpersonal realities. Our human fulfillment and well-being seem essentially to require relation to others, particularly in such goods as friendship, marriage, and family. To narrow our reproductive world in this self-focused way, by becoming autonomous procreators, seems reductive of our possibilities for well-being, by removing us from those others with whom we would have had the potential for interpersonal communion. Cloning is akin to masturbating for children, a lonely and solipsistic business; a world in which all reproduction had become disjoined from sex would be similarly depersonalized, as Kurzweil's visions of virtual sex between avatars makes clear.

Second, this reduction is especially apparent in the relationship to the created children that are the fruit of a technical process, rather than of a loving, human, relationship. Children created through technical means are ultimately, in the eyes of their creators, artifacts, products of their will and design, rather than persons. Parents-to-be hope for children, rather than plan them (they can, of course, plan *for* them); cloners will be the designers and manufacturers of their children. Cloning parents, by adopting this relationship, alienate themselves from their children. The relationship of cloners to their children is stunted as well, leaving again a lesser scope in the relationship to basic goods. What would be true in cloning would be true on a greater scale were the envisaged changes in human reproduction to become part of our nature.

Such considerations form the groundwork for arguments against research aimed at limiting our species' dependence on sexual reproduction as such; the state of affairs envisaged in such research is one intrinsically limiting of the goods available to us in action. In so limiting these goods, the envisaged future, in which reproduction is severed from sex, constitutes a genuine threat to our well-being, a threat most vividly seen in imaginative thought experiments, such as that of Aldous Huxley in *Brave New World*.

A related threat to one of the basic goods is also welcomed as desirable by some futurists such as Kurzweil. An important fact about both our practical and our animal natures is the separateness of persons. No person is a part of another, and there are definite limits to the extent that persons can share in experiences, choices, and actions. Now, on the one hand, such boundaries seem to limit the extent to which certain goods are available to us, such as friendship. Yet, on the other hand, these boundaries are also necessary for that good. For in the absence of boundaries, it is impossible to make the choices to share, to communicate, to give of one's self to another (and to withhold from yet others), that are necessary to make relationships of any sort possible. Recognition of this circumstance was at the core of Aristotle's rejection of Plato's politics, which in the *Republic* involved a wholesale attack on personal property and privacy.

Yet Kurzweil, for example, anticipates a merger of biological and technological intelligence that will essentially break down such barriers which, he says

> . . .are just a limitation of biological intelligence. The unbridgeable distinction of biological intelligence is not a plus. "Silicon" intelligence can have it both ways. Computers don't need to pool their intelligence and resources. They can remain "individuals" if they wish. Silicon intelligence can even have it both ways by merging and retaining individuality—at the same time. As humans, we try to merge with others also, but our ability to accomplish this is fleeting(Kurzweil, 2005, p. 376).

The merger of biological and silicon intelligences that Kurzweil foresees here seems to me to threaten the conditions under which friendship at large is possible, conditions that might seem like "limitations," but are essential to the basic human good of friendship.

The second threat mentioned above was also to our practical nature, namely, the threat against our abilities to participate actively in our own well being, in the making of our own lives and choices. Various enhancers currently pose just such a risk when taken, not to alleviate a condition that itself poses a threat to our activity, but to substitute for the labor of making ourselves more disciplined, harder working, more in control of our emotions, more capable in our relationships to others, more able to hit a baseball or run a race, and so on. The use of Prozac and Ritalin, not by those who are sick, but by those who simply wish to perform better, or the use of steroids by athletes, among other things, makes us increasingly passive in our pursuit of well being (The President's Council, 2003).[9] The threat here may be brought out by a brief account of Robert Nozick's famous "experience machine" thought experiment.

Nozick's thought experiment centers on a machine that promises the experience of a fully meaningful life (Nozick, 1974, p. 43). Plug yourself into the machine, and you will be provided the experience of creating great art, pursuing valuable relationships, and thinking great thoughts. Nozick's intuition, shared by most philosophers, is that plugging in would be supremely undesirable, regardless of what form of life we would therein experience. For ultimately, what the machine offers, with regard to any of the various aspects of a meaningful life, is mere appearances; but what human beings desire, what matters, is what is available to us through genuine action.

What characterizes the difference between the experience of some activity and the reality? Some capacity of the human being must be actualized in genuine action that is not actualized in the mere experience of performing that action. In the experience machine, it seems that what is missing is any act of the will. On Nozick's own understanding, the experience machine renders us utterly passive. We are unable to act, while the machine acts on us. Since there is an experiential aspect even to acts of the will, the machine will, in the course of providing us with the experience of a meaningful life, provide us also with the experience of genuine activity. But confronted with the experience machine, we distinguish between willing something and the experience of willing something, performing an action, and experiencing that performance. What emerges is that experience that is not grounded in the reality provides only the illusion of the truly desired reality (Finnis, 1984; Tollefsen, 2003).[10]

The experience machine should help us to get a handle on what is disturbing about the use of Ritalin or Prozac in the healthy. To the extent that, as Nozick's thought example revealed, our active self-constitution is what we want, rather than merely external satisfactions, or successes, use of drugs and other enhancers to accomplish what we should accomplish through our own labor is actually unenhancing.

Two caveats are necessary here. First, many of these enhancers have other legitimate uses—to treat depression or hyperactivity, or to provide therapy to those suffering from muscular disorders, for example. Here is a good instance of a case in which, although potential negative side effects exist, and might have been, and perhaps were, foreseen, the fact of those negative side effects alone does not seem conclusive evidence that the research that brought about these developments should never have been carried out. By the same token, if there are drugs or techniques being developed which serve no such therapeutic purpose, but would only serve simultaneously to bring us greater success in some area, while making us more passive in respect of that success, then research into these types of enhancement seems threatening to our actively practical nature.

An obvious objection to this raises the second important caveat, however. Ritalin, Prozac, memory aids, steroids, and so on, are far from removing every aspect of our control of our lives, even in some limited domain, from us. Although the healthy student who takes Ritalin to do better on his exams does both provide himself with an advantage and, as it were, borrow that advantage from an external source, it is still the case that whatever knowledge is displayed by the student is largely the student's own. He has not completely surrendered his active participation in his own self-constitution.

There are, however, aspirations on the part of some to more or less completely bypass such activity. It will be helpful to divide the discussion at this point into a consideration of two different sorts of goods in relation to which we might become artificially passive.

Consider, for example, drugs or surgical techniques that could be developed which would make us professional class athletes without the need for workouts, diets, and weight-lifting, or memory downloads that would provide us with knowledge of places we had never been, or fields we had never studied. There can be little doubt that the agents who took such drugs or received such implants would in fact be healthier and more knowledgeable than those who had not. So far forth, such agents would indeed be better off than their less advantaged peers. Life and knowledge are substantive goods—our participation in such goods contributes to our well-being even apart from our having chosen so to participate.

By contrast, consider the good of friendship. Suppose that technologies became available that would create in two people the belief that they were friends, or would virtually create a world for an agent in which she was happily married. Such technologies are variants on Nozick's experience machine. Similar technologies would enable us to download virtues, or even biographies, constructed according to our desires. If we wished to be braver or more temperate, we could download a character change, becoming virtuous without pain. Who could object to such a manner of becoming better? Perhaps we dislike not simply our character, but the entire life that has brought it to be as it is. We could then download a new identity, in the so-called narrative sense, a new story-history about ourselves that we would now take to be real. Or, perhaps we are unhappy with our integrity: our feelings, choices, and reasons for action might often fail to cohere. Reconfiguration of our identity might solve precisely this problem (see Kurzweil, 1999, 2005).[11]

Such scenarios seem farfetched. But, on the one hand, they seem to be variations of the images provided by some proponents of new genetic technologies to rid the world of crime by genetically making all children "good," or more social and altruistic, less aggressive and self-centered. And, on the other hand, while there seems little reason to think such drugs, techniques, and so on are around the corner, there is certainly research oriented to the discovery of precisely these sorts of technologies, techniques, drugs and so on. Perhaps such research is a colossal waste of time. Is it also morally suspect?

What we should note about goods such as friendship, virtue, good character, integrity, and the like, is that all require, in order to be genuine, acts of the will. True friendship cannot be downloaded, uploaded, or installed, for true friends make mutual commitments to one another's goods, and no genuine commitment can be made by anyone except the agents in question. Two people may be forced to marry, but they can never be made to love one another. So unlike the cases involving substantive goods like knowledge or health, when techniques or drugs become available that will make us more virtuous, that will create virtual friends, that will eliminate crime, will create new biographies for us, and so on, these techniques will in no sense make us better off with respect to the goods being sought. They will, indeed, bypass those goods altogether by removing a necessary condition for their reality, human choices, and commitments.

Any such "enhancement" thus poses a very serious threat to our well-being and to our practical nature. Nor do there seem to be any legitimate uses of such technologies that would not involve exactly the same "moral bypass," a side run round our ability practically to constitute our own selves. Such technologies would therefore, I conclude, be morally wrong. So all research aimed at discovering ways to download character traits, biographies, or to genetically eliminate crime or hostility, insofar as these are not unwilled consequences of pathologies, should likewise be viewed as morally misguided. The prospect of a world in which everyone acts morally is only a promising prospect if it is genuinely a world in which everyone acts.

What, though, about the substantive goods, which really do make us better off even when we participate in them through no choice of our own? Here too, I think, we should see a significant value to our own self-constituting activity. Although it is good to be healthy, or knowledgeable, whether we have actively pursued health or knowledge or not, we should find it more desirable to have become healthy or knowledgeable through our own activity. Health and knowledge do not simply benefit us in the obvious way, namely by making us healthy and knowledgeable, they also help to define who we are. The body-builder or long distance trainer has defined herself in a certain way as a healthy person; the philosopher has similarly defined herself as a seeker of wisdom. Such self-definition as a person for whom one or another of the goods are significant is possible only because we choose actively to pursue these goods, to cultivate them, and to own our participation in them. All this would be lost if there were a pill for bodily strength, or wisdom, or ability, and so on.

Here, though, we can see possible therapeutic uses to drugs or surgical techniques aimed at health or perhaps knowledge. Bodies and minds grow ill and decay, and some of this sickness and decay is premature. It is reasonable to hope for cures for muscle wastage or Alzheimer's disease, but critical to remember the importance of self-constitution even here. While research in such areas cannot be ruled out, as it can where reflexive goods are concerned, still there must be concern for the uses to which new enhancers are put, and a determination to steer research as much as possible to the genuinely therapeutic. As a general rule, I would suggest, such research stays on safer ground insofar as it is concerned with the protection and restoration of our *capacities*, rather than with providing us directly with that which someone with functioning capacities and an active will could achieve.

Finally, what of changes to our bodily identity? Can lines be drawn here to protect our animal identity? Two points should be clear straight off. First, some changes in our bodies would be such as to, again, rupture our relationship to certain basic goods. So if, pursuant to the goal of making us autonomously reproductive, changes were systematically introduced in the human body to make us, as a species, or as individuals, increasingly asexual, i.e., no longer differentiated sexually as male and female, this would radically rupture our relationship to the goods of marriage and procreation. Or, suppose that our bodily senses were gradually eliminated in favor of computer navigation systems implanted in our brain with some sort of device for environmental input. One purpose of this might be to increase our ability to respond in common to certain stimuli, for example, among soldiers in war (Hughes, 2004).[12] But we should ask ourselves, would this radically diminish our capacity for enjoyment of the arts of music or painting? To sacrifice these capacities for new abilities of a purely instrumental value would seem a violation of the goods of the person.

Second, some bodily changes could no doubt make it more difficult for us to be self-constituting agents. While the ability to make choices is not a bodily-based ability—no mere animal could make choices, but only a spiritual animal—it is an ability deeply conditioned by the existence and experience of our animal body. Changes in the brain, or perhaps even other parts of the body, might erode this necessary condition of freedom and reason and render us less able to act autonomously, and in certain respects, in a more merely animal manner. Imagine, for example, a change that made us *more* susceptible to physical pleasure than we are already. Such a change might be considered attractive in our hedonistic society (where men take Viagra not to cure impotence, but to make sexual intercourse more pleasurable). Such a change would make temperate choices more difficult, and would reduce our scope for truly human action.

But these threats—to our relationship to certain goods, and to our ability to judge and choose—I would argue, are the *only* real threats to our human identity. Insofar as bodily changes could be brought about that would not negatively bear either on our relationship to the basic goods, or our ability to actively pursue them, such changes would not constitute *any* genuine change in human nature. For our human nature is determined by our being animals for whom *this* precise list of goods is constitutive of well-being, which well-being can be pursued in action following free and deliberate choice. So long as we remain animals, benefited by these goods, and capable of constituting our selves through the pursuit of these goods, we remain the same species.

Suppose that, not by artificial means, but by the pressures of evolution, we grew a third leg. Would this constitute a change in human nature? It would not. Nor, it seems, would the development of some new organ, or the enhancement of some organ, in such a way as radically to change our physical abilities: if we grew gills, or something analogous, and were able to stay under water for hours at a stretch, this would not constitute a change in our nature. Moreover, some such changes might genuinely result in new opportunities to pursue human goods in creative ways.

If natural selection could change us without essentially changing our natures, would it therefore be morally permissible deliberately to bring about such changes in our animal bodies? Is research into at least some such possibilities morally permissible precisely because it would not be impermissible to utilize the fruits of such research? Surprising though it may seem, I think such research might, within certain bounds, be acceptable. Moreover, various positive advantages can be envisaged to many conceivable changes—advantages to health, for example, or to our perceptual abilities.

However, there is one final point to be made about the bearing of our human nature on research into human enhancement. Consider again the claim that what we are essentially is human beings, organisms of the human species. What bearing does this have on questions of human rights? Some rights accrue to individuals in virtue of some contingent feature or property they possess; the right of a student to have work returned promptly, for example, and to have her work graded fairly.

Other rights, however, accrue to individuals in virtue of what they are: our basic human rights are not thought to vary in accordance with accidental properties, but to belong to us for as long as we exist. If what we are, essentially, is a human being, then we are subjects of basic rights—such as the right to life—from the moment we come into existence as human beings until the moment we cease to exist as human beings.

Despite arguments to the contrary, it does not seem plausible to suppose that there are reasonable markers for our beginnings and endings other than conception, and the permanent end to our unified functioning as an organism, i.e., to biological death. We come into being (unless we are monozygotic twins) when a single cell organism begins to exist, genetically distinct from its parents, and capable of its own self-directed growth and maintenance. We cease to exist at biological death. We are subjects of basic rights for the duration of this bodily existence.

Clearly, some, at least, of the research that is envisaged as necessary to make possible enhancements to human nature will require experimentation on early human embryos. Of the three planks of the predicted revolution in human nature, genetics certainly, and possibly both nanotechnology and robotics, will be wholly successful only if their integration into our current biological existence is furthered by knowledge primarily available through human embryonic research. If we are human beings, who begin as embryos, then such research, much of which is destructive of the embryo, and all of which, clearly, takes place without the embryo's consent, is immoral.

This moral boundary is contingent; it might well become possible to rely on computer models and animal experimentation to obtain the knowledge we seek, without the necessity of invasive and destructive embryo experimentation. But for the moment, the boundary seems such a considerable hedge around research into changes to our animal bodies that this sort of research should be considered morally impossible. The singularity, if we are morally responsible, might not be so near after all.

The boundaries are those already established in this chapter: no research that involves damage or destruction to human persons, including human embryos; no research oriented towards drugs or techniques that would change our essential relationship to the finite set of basic goods, or that would reduce our ability actively to participate in those goods. And of course, no research that would violate informed consent. These boundaries might, however, constitute such a considerable hedge around research into changes in our animal bodies that, for all intents and purposes, such research should be considered morally impossible.

Notes

1. GNR stands for genetics, nanotechnology and robotics, of which artificial intelligence in a branch.
2. See Naam Ramez (2004); also see the various statements and documents available at www.transhumanism.org.
3. There are responses to such possibilities that blur the lines between arguments about negative consequences, and concerns about changing human nature. Fukuyama believes that altering human nature by, e.g., eliminating aggression or alcoholism would likely have negative "flattening" consequences in other aspects of our nature, making us less creative, for example, or social or cooperative.
4. See, for discussion of such possibilities, the papers in Mihail C. Roco and William Sims Bainbridge (2002).
5. The view I describe and defend in this section has come to be known as "Animalism." The central texts in the recent work defending it include Eric Olson (1997a, b), David Wiggins (1980), and Paul Snowdon (1990).

6. For further reflections along these lines, and an outstanding attempt to present a unified view of the human person as both animal and spirit, see Braine (1992).
7. The view I articulate here—a goods-based version of natural law theory—was earliest, and, in my view, most successfully presented in the twentieth century, by Grisez, Boyle, and Finnis. See Grisez (1965, pp. 168–201), Grisez et al. (1987, pp. 99–151), and Finnis (1980). The Grisez-Boyle-Finnis approach has been followed especially by Robert P. George (1999). More recently, and in slightly differing ways, a goods-based view has been developed by T.D.J. Chappell (1997), David Oderberg (2000), Mark Murphy (2001), and Alfonso Gomez-Lobo (2002).
8. See, for example, the wide array of extramarital and extra-sexual forms of reproduction contemplated in Silver's *Remaking Eden*.
9. See the related discussion in The President's Council (2003), Chapters 3 and 6.
10. See the discussions of this in John Finnis (1984, pp. 37–42) and Christopher Tollefsen (2003, pp. 153–164).
11. See Ray Kurzweil (1999, 2005).
12. "The Defense Advanced Research Projects Administration is providing tens of millions of dollars to MIT, Duke and other universities through their 'Brain Machine Interfaces Program' aimed at permitting soldiers to communicate with equipment, and each other, at the speed of thought. A lot of research is focused on creating new materials that can coexist with neurons indefinitely, and can be made into nanoscale electrodes" (Hughes, 2004, p. 40).

Bibliography

Braine, D. (1992). *The human person*. Notre Dame: University of Notre Dame Press.
Chappell, T.D.J. (1997). *Understanding human goods*. Edinburgh: Edinburgh University Press.
Finnis, J. (1980). *Natural law and natural rights*. Oxford: The Clarendon Press.
Finnis, J. (1984). *Fundamentals of ethics*. Washington, DC: Georgetown University Press.
Fukuyama, F. (2002). *Our posthuman future: Consequences of the biotechnology revolution*. New York: Farrar, Straus and Giroux.
George, R.P. (1999). *In defense of natural law*. Oxford: Oxford University Press.
Gomez-Lobo, Alfonso (2002). *Morality and the human goods: An introduction to natural law ethics*. Washington, DC: Georgetown University Press.
Grisez, G. (1965). The first principle of practical reason: A commentary on the summa theologiae, I–II, question 94, article 2. *Natural Law Forum*, 10, 168–201.
Grisez, G., Boyle, J., and Finnis, J. (1987). Practical principles, moral truth, and ultimate ends. *American Journal of Jurisprudence*, 32, 99–151.
Hughes, J.H. (2004). *Citizen cyborg: Why democratic societies must respond to the redesigned human of the future*. Cambridge: Westview Press.
Huxley, A. (1998). *Brave new world*. New York: Perennial Books.
Krauthammer, C. (2001). Why pro-lifers are missing the point: The debate over fetal-tissue research overlooks the big issue. *Time*, February 12, 2001, 60.
Kurzweil, R. (1999). *The age of spiritual machines: When computers exceed human intelligence*. New York: Viking.
Kurzweil, R. (2005). *The singularity is near: When humans transcend biology*. New York: The Penguin Group.
Oderberg, D. (2000). *Moral theory: A nonconsequentialist approach*. Oxford: Blackwell Press.
Olson, E. (1997a). *The human animal: Personal identity without psychology*. Oxford: Oxford University Press.
Olson, E. (1997b). Was I ever a fetus? *Philosophy and Phenomenological Research*, 57, 95–110.
Murphy, M. (2001). *Natural law and practical rationality*. Cambridge: Cambridge University Press.
Nozick, R. (1974). *Anarchy, state, and utopia*. Basic Books: New York.

Ramez, N. (2004). *More than human: How biotechnology is transforming us and why we should embrace it.* New York: Random House.
Roco, M.C., and Bainbridge, W.S. (eds.) (2002). *Converging technologies for improving human performance.* Washington, DC: National Science Foundation. [On-line]. Available: www.wtec.org/ConvergingTechnologies
Silver, L. (1998). *Remaking eden: Cloning and beyond in a brave new world.* New York: Avon.
Snowdon, P. (1990). Persons, animals, and ourselves, in C. Gill (ed.), *The person and the human mind: Issues in ancient and modern philosophy*. Oxford: Oxford University Press.
The President's Council on Bioethics. (2003). *Beyond therapy: Biotechnology and the pursuit of happiness*. Washington, DC: U.S. Government Printing Office. [On-line]. Avaliable: www.bioethics.gov/reports/beyondtherapy
Tollefsen, C. (2003). Experience machines, dreams, and what matters. *Journal of Value Inquiry,* 37, 153–164.
Tollefsen, C. (2006). Persons in time. *American Catholic Philosophical Quarterly,* 80, 107–123.
Wiggins, D. (1980). *Sameness and substance*. Oxford: Basil Blackwell.

Synderesis, Law, and Virtue

Angela McKay

Recent scholars have made much of the fact that natural law and prudence are not independent. Natural law, these scholars argue, cannot be properly applied without the virtue of prudence. Indeed, some of these scholars argue, prudence is in a sense "prior" to the natural law (Nelson, 1992). Others, taking a more nuanced approach, have pointed to the interdependence of prudence and the natural law (see, e.g., Hall, 1994). Although I think it is incorrect to claim that prudence is "prior" to the natural law, I believe that scholars are right to emphasize the important role that prudence plays in the application of the natural law, and in this paper I wish to explore the connection between not only prudence, but moral virtue in general and the natural law. Insofar as I agree with these scholars that the natural law cannot be properly applied without prudence, this paper will add little to the ground that scholars such as Pamela Hall have already prepared. In this paper, however, I wish to push the connection between law and moral virtue even further by exploring their common source: synderesis. For while many tend to think of synderesis in conjunction with man's natural habitual knowledge of the first principles of natural law, a less explored aspect of synderesis is the role that synderesis plays as the source of moral virtue.[1] An understanding of this role and of the subsequent role of virtue in the development of man's ability to apply synderesis in action, I will argue, allows us to better understand the intimate connection between Aquinas's respective theories of virtue and the natural law.

In the first part of this paper, I will explore the Aristotelian concept of nature that lies at the heart of Aquinas's theory of the acquired virtues.[2] For Aquinas as for Aristotle, nature provides things not merely with the end that they are ordered to, but with the very principles that allow them to arrive at those ends. In the second part of this paper, I will examine the "principles" that, according to Aquinas, allow man to acquire the virtues that order him to the good commensurate with his nature; i.e. the good of reason. Foremost amongst these principles, I will argue, is synderesis. Synderesis, together with man's natural habitual knowledge of the first principles

A. McKay (✉)
Department of Philosophy, Catholic University of America, Washington, DC, USA
e-mail: mckay@cua.edu

M.J. Cherry (ed.), *The Normativity of the Natural,* Philosophical Studies in Contemporary Culture 16, DOI 10.1007/978-90-481-2301-8_3,
© Springer Science+Business Media B.V. 2009

of speculative reason, provides the foundation out of which the virtues arise. The virtues, however, emerge from this account as something that are as important, if not more important, than the principles that give rise to them. If the synderesis allows the moral virtues to come into being, it is moral virtue that allows synderesis to yield fruit. I will conclude this paper by addressing the question of what the intimate connection between the moral virtues and synderesis means for our understanding of Aquinas's theory of natural law.

1 Nature, Habit, and Virtue

Certain aspects of Aquinas's theory of virtue are well known, so well known, in fact, that it seems pointless to reiterate them. At first glance, the relationship between virtue and nature appears to be one such aspect. What I wish to examine here, however, is a portion of the relationship between nature and virtue that tends, I think, to receive less attention than it ought, namely the sense in which nature provides not merely the end to which a thing is ordered, but the very sources from which the virtues that order a thing to its end arise.

The central and most basic theme in Aristotle's view of nature, and one which will be central to Aquinas's account of the virtues, is the notion that a thing's nature not only provides the end to which a thing is directed, but also the sources from which a thing achieves its telos: things that exist by nature have interior principles ordering them towards their ends. In his *Physics*, Aristotle distinguishes those things that exist by nature from those things that are artifacts precisely on the basis of this distinction: those things that exist "by nature" contain something intrinsic to themselves that causes them to be what they are, and which drives those things towards the realization of their *telos* (Aristotle, *Physics II*, Chapter 1). A thing's nature, its "interior principle of change and rest," affects a thing's development in the first and most obvious way by providing an end towards which the thing in question is directed. To understand, for instance, the nature of a frog, is to understand that there are certain activities which capture what it means to be a frog. Even our understanding of "tadpole" is informed by our understanding of "frog," for to understand what it is to be a tadpole is to understand it as something that is on its way to becoming a frog; something that is meant to do the sorts of things that frogs typically do. Thus in one very important sense a thing's nature is its end, or telos: "the nature is the end or 'that for the sake of which.' For if a thing undergoes a continuous change and there is a stage which is last, this stage is the end or 'that for the sake of which' " (Aristotle, *Physics II*, Chapter 2). While the provision of ends is certainly a fundamental component of the idea of nature, it is hardly the only one.

Nature directs things to their ends in a second, less obvious way, insofar as it provides things with the very principles that direct them to their ends. To truly understand what it is for something to have a nature, one must understand nature not only in terms of a thing's end, but in terms of nature's ability to direct things to their ends. That is, we have to see that the nature of the frog or the oak is already present in the tadpole or acorn as a principle driving it towards its completion. The tadpole is already in some sense a frog, for it is on the way towards becoming one, and that it is, is due to its nature. Nature thus provides both the source of motion towards the end and the end itself. As such, nature is present in every aspect of a thing's development.[3]

The Aristotelian notion of nature as providing both the end to which a thing is ordered and the principles that drive the thing towards the fulfillment of its nature, I will argue, have a fundamental importance for understanding the connection between virtue and natural law in Aquinas. For, as I shall show in what follows, whenever Aquinas wishes to give an account of a given type of virtue, he is attentive to both (a) the "seeds" or principles from which those virtues spring and (b) the "rule" or measure of those virtues—i.e., what it is those virtues are ordered towards. Good habits, or virtues, take the first, general principles that orient man towards his end and enable him to act according to those principles in his particular actions. The rule or measure of those habits is of course whether or not they order man to his end. Importantly, however, habits and virtues arise *from* something, and it is in exploring this something that we begin to see the centrality of the role that synderesis plays in Aquinas's account of the acquired virtues.

2 The Seeds of Acquired Virtue

In what follows, I wish to examine Aquinas's account of the "principles" from which the acquired virtues arise. An examination of these principles, I will argue, indicates that the primary source of the moral virtues is the natural habitual knowledge of the first principles of action, or synderesis. In both the *prima secundae* and the *DQV*, Aquinas's discussion of acquired virtue is preceded by a discussion of whether or not man possesses any virtue "by nature."[4] In the course of explaining whether and to what extent man possesses virtue "by nature," Aquinas describes certain natural active principles that he refers to as the "seeds" of the acquired virtues. These "seeds," as we shall see in what follows, are the naturally known principles of thought and action. It is man's natural knowledge of these principles that gives rise to his desire for the good of reason, and these principles, coupled with man's desire for the good of reason, are the foundation upon which the acquired intellectual and moral virtues rest.

Aquinas responds to the question of whether man possesses any virtue "by nature" with a qualified "yes." Although man does not naturally possess the virtues in their completion, he does naturally possess an "aptitude" for virtue, both insofar as he is naturally capable of receiving virtues and naturally capable of creating them, and he possesses this aptitude to some extent in virtue of both his specific and his individual nature. In the process of delineating these various aptitudes, Aquinas describes a specific kind of natural habitual knowledge that, he claims, is the "seed" of all the virtues that man subsequently acquires.

After rejecting both the possibility that the virtues are innate and merely uncovered and the position that the virtues are bestowed on man entirely from without, Aquinas opts for a middle position, namely the position that "science and virtue are in us by according to aptitude, but not according to perfection" (*ST* I-II q.63 a.1; see also *DQV* a.8 and *DV* q.11 a.1). Virtue—or at least, acquired virtue—will thus be for Aquinas something that both is and is not natural to man.

In the *DQV*, Aquinas proceeds to distinguish two different ways in which man can possess an "aptitude" for virtue. A thing can possess an "aptitude" for a form in a purely passive sense, insofar as it is capable of receiving a form, but incapable of creating that form in itself. This is the sense, Aquinas explains, in which air has an "aptitude" for receiving the form of fire (*DQV* q.1 a.8). On the other hand, a thing can possess an "aptitude" in *both* an active and a passive sense, in the way that the body is capable of receiving the form of health (*DQV* q.1 a.8). As opposed to those things that merely possess a passive potency for a form, those things that possess *both* an active and a passive potency are capable of causing that form in themselves (*DQV* q.1 a.8). This latter type of aptitude, Aquinas continues, is the type of aptitude for virtue that all men naturally possess. This aptitude, moreover, is present in virtue of both man's specific and his individual nature. Although Aquinas proceeds to describe the natural active and passive potencies for virtue that man possesses in virtue of his specific and individual nature, however, it soon becomes clear that man's active potency for virtue is present only in virtue of his specific nature, and moreover, only in two of the three faculties of the soul.

After stating that man naturally possesses both an active and a passive capacity for virtue, Aquinas appears to indicate that all three of the faculties of the soul possess both an active and a passive capacity for virtue. When Aquinas proceeds to describe the kind of receptivity that each faculty of the soul possesses, however, it becomes clear that only the intellect and the will possess both active and passive capacities for virtue. Because the intellect is capable of receiving intelligible things, and because intellectual virtue consists in the knowledge of intellectual things, it is clear that the intellect has a passive capacity for virtue (*DQV* a.8). The intellect, however, also has an active capacity for virtue, that is to say, the ability to create virtue in itself. This occurs insofar as man also possesses active intelligence, since it is through the light of reason that things become intelligible in the first place (*DQV* a.8). The important point for our purposes, however, is that Aquinas also believes that man needs more than *merely* the ability to receive intelligible things and the light of reason: reasoning at both the practical and theoretical level must *begin* from somewhere; there must be some foundation for reasoning to rest upon. Thus in addition to the natural light of reason, Aquinas holds that all men also possess natural habitual knowledge of the first principles of practical and theoretical reason (*DQV* a.8). Because all reasoning about theoretical and speculative matters takes its point of departure from these first principles, Aquinas says that these principles are the "seeds" of the intellectual and moral virtues: all the practical and theoretical knowledge that man gains as he accumulates the virtues derives from these naturally known first principles.

Like the intellect, Aquinas argues that the will also naturally contains both the active and passive capacity for virtue. Insofar as all men naturally desire the good of reason (*ST* I-II q.63 a.1; *DQV* a.8.),[5] men possess a natural active principle inclining them towards virtue, since this inclination causes men to seek the good of reason (*DQV* q.1 a.8). However, since the will can be inclined in many different ways in its pursuit of the good, it is receptive to dispositions that order it in a specific direction, and hence also naturally contains a passive capacity for virtue (*DQV* q.1 a.8).

Although Aquinas initially claims that all three faculties of the soul have both an active and a passive capacity for virtue, at least in some respect, it is noteworthy that he does not describe an active capacity for virtue in the "lower desire," the third faculty of the soul. To the contrary, Aquinas merely says that because the concupiscible and irascible faculties are naturally capable of listening to reason, they are "naturally receptive to virtue, which is perfected in them insofar as they are disposed to follow the good proposed by reason" (*DQV* q.1 a.8). Man's specific nature, then, enables him to create virtue in himself, not merely because the three faculties of the soul are each receptive to virtue, but also because (a) the will naturally desires the good of reason, (b) the intellect possesses natural habitual knowledge of the first principles of thought and action, and finally (c) man naturally possesses the natural light of reason. Together, these three active principles are able to create habits in the faculties of the soul.

The three principles mentioned above will be explored in more detail in what follows, but first it is appropriate to examine man's individual nature, for Aquinas also maintains that virtue is "in some sense" natural to man's individual nature. What will emerge from Aquinas's account of the "sense" in which virtue is natural to man's individual nature, however, is that man's individual nature plays far less of a role in the generation of virtue than man's specific nature does.

Aquinas holds that man is naturally disposed to virtue on the part of his individual nature insofar as man's specific bodily makeup makes him more inclined to some acts of virtue than to others (*DQV* q.1 a.8). Some men, for instance, are naturally less fearful than others. Those who are naturally less fearful, of course, will find it easier to perform courageous actions than those who lack such a natural inclination. Since such inclinations lack the guidance of reason, they can no more be considered "full" virtues than the natural predispositions towards virtue that man possesses in virtue of his specific nature. It is worth noting, however, that although Aquinas says that there can be an "aptitude" for virtue on the part of both man's specific and his individual nature, he also indicates that the aptitudes natural to man's individual nature are not, properly speaking, *sources* of the virtues. A natural bodily inclination, Aquinas notes, may dispose man towards the act of a certain kind of virtue and hence is a "sort of" predisposition towards virtue, but it does not incline man towards the acts of all the virtues. To the contrary, insofar as man has a bodily inclination towards the activity of one virtue, he also has an inclination *against* the activity of another virtue: "Those things which are dispositions to all the virtues, are not able to be by natural inclination; because a natural disposition which inclines to one virtue, inclines to the contrary of another virtue" (*DQV* a.8 ro.10). Those who have a natural bodily inclination towards courage, for example, are thereby also less naturally inclined towards patience, since they have more difficulty restraining the irascible appetite, and patience requires that the irascible appetite be restrained (*DQV* a.8 ro.10). Evidence of this, Aquinas notes, can be found in animals: "Whence we see that an animal which is naturally inclined towards the act of one virtue, is naturally inclined towards the contrary defect of some other virtue; as the lion which is naturally brave is also naturally cruel" (*DQV* a.8 ro.10). In order to achieve the good of virtue man thus needs dispositions beyond his bodily inclinations, namely inclinations that dispose him towards acts of all the virtues. These, says Aquinas, are found in the reason, and the natural principles that are found there are the seeds of all the virtues: "But men are intended to achieve the full good of virtue; and thus it is fitting that they should possess an inclination to all acts of virtue: and because it cannot come from nature, it is fitting that it should come from reason, in which exists the seeds of all the virtues" (*DQV* a.8 ro.10). The only real candidates for the "seeds" of the virtues, then, are the general principles of thought and action that man possesses in virtue of his specific nature.

On the basis of Aquinas's discussion in article 8 of the *DQV*, then, it seems as if the natural active principles that allow man to create virtue in himself are (a) his natural habitual knowledge of the first principles of thought and action, or the "seeds" of the virtues, (b) the will's natural affection for the good of reason, and (c) the "light of reason." The last of these, as we shall see in subsequent sections, is what enables man to render the first of these specific in action. However, before turning to this, we should say something about the relationship between (a) and (b). An examination of this relationship, I will argue in what follows, reveals that the two primary active principles that allow man to create the acquired virtues in himself are (a) and (c).

3 Natural Knowledge as the Source of Natural Desire

In his account of faith in the *De Veritatae*, Aquinas indicates that it is man's natural knowledge of the first principles of thought and action that gives rise to his natural desire for the good of reason. Although Aquinas only makes this point in order to draw a comparison with the role of faith at the supernatural level, the explanation that he offers here allows us to add an important clarification to his account of the natural "active" principles that allow man to acquire virtue.

In the second article of question 14, Aquinas explores a certain definition of faith, namely whether faith is "the substance of things hoped for, the evidence of things that appear not" (*DV* q.14 a.2). Aquinas proceeds to defend this definition on the grounds that "the intellect is determined to something through the will, and the will does nothing except insofar as it is moved by its object, which is the good to be sought for and its end. In view of this, faith needs a two-fold principle, a first which is the good which moves the will, and a second which is that to which the understanding gives assent under the influence of the will" (*DV* q.14 a.2). Then, in a move which—at least for Aquinas—is highly typical, Aquinas points out that the "good" that moves the will "is twofold": "Man, however, has a two-fold final good, which first moves the will as a final end" (*DV* q.14 a.2). The first of these two goods "is the happiness of which the philosophers speak," namely that good proportionate to human nature, which man is capable of achieving through his natural powers (*DV* q.14 a.2). The second, however "is out of all proportion with man's nature because his natural powers are not enough to attain it either in thought or desire" (*DV* q.14 a.2). Since it will be because of this latter good that man gives assent to the things he holds by faith, Aquinas proceeds to contrast the way in which man is ordered to the good of reason with the manner in which man is ordered towards the good of supernatural beatitude. The first half of this explanation is relevant for our purposes insofar as it explains how man's desire for the good of reason arises.

Aquinas's explanation of the origins of man's desire for the good of reason begins by appealing to the feature of nature that we discussed earlier in this chapter: nature does not only provide things with their ends; it also provides man with the very principles that order him to that end: "Nothing can be directed to any end unless there pre-exists in it a certain proportion to that end, and it is from this that the desire of the end arises in it" (*DV* q.14 a.2). This "proportion" is achieved, Aquinas proceeds to argue, "insofar as, in a certain sense, the end is made to exist inchoatively within it, because it desires nothing except insofar as it has some likeness to the end" (*DV* q.14 a.2). Man's end is made to exist "inchoatively" within him, moreover, insofar as he possesses the "seeds" of the intellectual and moral virtues: "This is why there is in human nature a certain initial participation of the good which is proportionate to that nature. For self-evident principles of demonstrations, which are seeds of the contemplations of wisdom, naturally pre-exist in that good, as do the principles of natural law, which are the seeds of the moral virtues" (*DV* q.14 a.2). The preceding quote is interesting for two reasons. First, it indicates that it is *because* man possesses the seeds of the intellectual and moral virtue that he desires the good of reason. This in and of itself allows us to draw a connection between the two natural active principles of virtue described above. Man is ordered towards virtue both insofar as he possesses natural habitual knowledge of the first principles of thought and action and insofar as he naturally desires the good of reason, but he naturally desires the good of reason because, thanks to his possession of these first principles, he has inchoate knowledge of that good.

This text is interesting in a second way insofar as it indicates that the naturally known principles of action are *both* the seeds of the moral virtues *and* synderesis, the natural habitual knowledge through which man knows the first principles of natural law. When we examine the descriptions of synderesis that Aquinas offers in other texts, it becomes clear that the "seeds" of the moral virtues Aquinas describes in the texts examined above and synderesis are one and the same thing. Synderesis, Aquinas tells us, is a habit that plays the same role at the level of practical action that the first principles of speculative reason play in the acquisition of knowledge (*ST* I q.79 a.12; I-II q.94 a.1). In some places, Aquinas even explicitly links synderesis and the "seeds" of the virtues, saying that together with the first speculative principles, synderesis is a "kind of seed plot containing in germ all the knowledge which follows," and in another place, tacitly accepts Augustine's claim that in man's natural power of judgment there are certain rules and seeds of virtue that we call synderesis (*DV* q.16 a.1; *ST* I q.79 a.12).

Synderesis, then, is in a very real sense the source of acquired moral virtues. Synderesis is importantly *not* moral virtue, because in itself it is insufficient to direct man in the multiplicity of decisions he must make. If this point is underemphasized in Aquinas's discussion of natural law, it certainly becomes apparent in the references to synderesis that occur in the context of Aquinas's discussion of the virtues. In both the *Prima Secundae* and the *DQV*, Aquinas ends his analysis of whether any virtue is "natural" to man by asserting that the natural principles that order man towards virtue have a very limited power. In order for virtue to be present, Aquinas insists, reason must be able to render these general directives specific in action: "In the same way that this predisposition to virtue without the activity of reason does not have the character of a complete virtue, neither do a few premises; for it is through rational inquiry that we reach specific conclusions from universal principles. It is also the job of reason to lead someone from his desire for the ultimate end to whatever is conducive to that end" (*DQV* a.8). In what follows, I wish to examine Aquinas's remarks about the process through which the seeds of the virtues give rise to virtues in their own right. Importantly, one only becomes able to reason correctly about the first principles of thought and action through developing the virtues.

4 The Rule of Reason

At a certain level, Aquinas's view of how the seeds of virtue give rise to the acquired virtues is too elementary to merit much discussion. In addition to the first, general principles of thought and action provided by nature, man is equipped with the natural light of reason and passions which are naturally receptive to formation by the dictates of reason. Thus through successive actions, based on reason's formulation of specific dictates from more general ones, man acquires the habits that dispose him to act in accord with the good of reason. It is important, however, that we spend some time discussing the importance of the role that reason plays in the acquisition of the acquired virtues.

In question 11 of his *De Veritate*, Aquinas considers the question of whether or not one man can "teach" another. Although Aquinas's primary concern in this article has to do with man's acquisition of scientific knowledge (*ST* I q.79 a.12), he notes that the same sort of explanation can be given regarding both the acquisition of virtue and the acquisition of knowledge. Since Aquinas's subsequent explanation is general enough to include both the formation of man's natural aptitude for scientific knowledge and man's natural aptitude for virtue, his explanation can be taken to apply equally to both.

After asserting that knowledge, like acquired virtue, comes neither exclusively from external sources nor entirely from within but partly from both, Aquinas offers an account of whether and how external agents can play a role in forming these "seminal principles". The explanation that he offers hinges on a distinction regarding the *kind* of potency for knowledge (and, by extension, acquired virtue) that man has. Some potencies, Aquinas notes, are "active" potencies. When something pre-exists in a thing as an active potency, the thing itself has the power to bring that potency into act (*DV* q.11 a.1). As an example of such an "active" potency, Aquinas points to the body's ability to heal itself (*DV* q.11 a.1). Other potencies, however, are "passive" potencies. When something pre-exists in a thing merely as a passive potency, then that potency can only be actualized by an external agent. This, says Aquinas, is what happens when air becomes fire, because the air has no power to create fire (*DV* q.11 a.1). Consequently, Aquinas will argue, when something pre-exists in a thing as an *active* potency, external agents can only assist in the actualization of the potency. This will be the case with the acquisition of both knowledge and virtue.

When something pre-exists as an active potency, external agents can only play an indirect role in the actualization of the potency. Specifically, although the external agent can provide a thing with the *means* of actualizing the potency, it is the thing itself that must actualize the potency (*DV* q.11 a.1). The doctor is said to heal the patient, but the doctor can only heal by providing nature with the means of healing itself. The doctor ministers to nature and assists nature, but it is nature that restores man to health.

Because knowledge and, by extension, acquired virtue, pre-exists in man as an active potency, the role of others in the acquisition of knowledge and acquired virtue must be similar to the role the doctor plays in healing the patient.[6] Just as with health, there are two ways of acquiring knowledge. One can either come to know something through one's own power or with the assistance of an instructor. In the latter case, however, the teacher's role is the same as that of the doctor's: the teacher can only minister to nature, in this case by helping man's reason (*DV* q.11 a.1). When the teacher teaches, that is to say, he must do so by appealing to the principles that the student already knows, and by helping the student to move from those principles to the appropriate conclusions. Were someone to propose things to a student in such a way that the student could not see how those things followed from first principles one would cause, says Aquinas, not knowledge but either opinion or faith (*DV* q.11 a.1). The important point here is that when a man learns, he—or more specifically his reason—and not the teacher or the book, is the primary cause of the acquisition. For even when man receives external assistance in the acquisition of knowledge—as he most often does—those external factors can help him only insofar as they appeal to his nature: his reason, and the first principles upon which that reasoning is based.

Given Aquinas's assertion that acquired virtue also pre-exists in man as an "active" potency, it is clear that a similar account has to be given regarding the acquisition of virtue. To say that the natural light of reason, in all men by nature, is the active principle behind the acquisition of virtue is not to say, as some might think, that our un-aided reason determines the correct act in every instance. It is to say, however, that even when we receive external assistance, that assistance merely ministers to nature, and it only ministers insofar as it helps man's reason to arrive at the appropriate conclusions. To acquire the virtues, therefore, simply is to learn to render the first, general principles of thought and action specific in the concrete instances of one's life.

One further point that arises from the above parallel between the acquisition of knowledge and the acquisition of virtue is the intimate involvement of the very rule by which a habit is measured in the formation of the habit itself. The rule, reason, is not merely an end, but is involved in the very acts through which the habits are formed. For it is reason that the teacher must appeal to in helping the student to find the appropriate path.

5 Synderesis, Law, and Virtue

Our discussion to this point, then, indicates that synderesis plays a pivotal role—perhaps, indeed, *the* pivotal role—in the acquisition of virtue. It is man's natural habitual knowledge of the first principles of thought and action that provides the explanatory force behind man's natural desire for the good of reason, and it is this natural habitual knowledge that provides the foundation for all further reasoning about moral matters. It is primarily this natural habitual knowledge, together with the natural light of reason, that enables man to create in himself the dispositions that order him towards the good of reason.

This background allows us to better understand an assertion recently put forward by some natural law scholars, namely the assertion that the natural law can not be properly applied without prudence. Those who emphasize the importance of prudence in the natural law typically point to the sixth article of question 47 in the *Secunda Secundae*, where Aquinas considers "whether prudence appoints an end for the moral virtues" (*ST* II-II q.47 a.6). When we read Aquinas's response, however, it is clear that in this article he merely reiterates the point made earlier in the *Prima Secundae*: synderesis is the "seed" of the moral virtues.

Aquinas addresses the question of whether prudence provides an end for the moral virtues by returning to the notion emphasized throughout his treatment of virtue, namely that practical and theoretical knowledge proceed from certain naturally known principles. The naturally known principles of practical reason, Aquinas continues, provide the ends of the moral virtues, because ends in matters of practice play a role similar to the role played by first principles in practical matters: "As in the theoretic reason there are certain naturally known principles, and some things naturally known through them, i.e. science, so in the practical reason there pre-exist certain naturally known principles, and these are the ends of the moral virtues, because the end in matters of practice operates like a first principle in matters of theory" (*ST* II-II q.47 a.6). Since it is synderesis that provides the ends of the moral virtues, Aquinas proceeds to argue that prudence does not provide ends, but merely helps us to reach the ends that synderesis provides. Prudence, in turn, helps the moral virtues achieve the ends set for them in advance by synderesis.

Thus, it is certainly true to point out, as these scholars do, that prudence and synderesis are interdependent. The preceding analysis, however, suggests a deeper point regarding why this is the case. Prudence and synderesis are interdependent, because synderesis and the moral virtues as a whole are interdependent. Synderesis provides man with the ground out of which the virtues can grow, but the virtues, in turn, enable man to apply the general knowledge that synderesis provides.

6 Natural Law and Virtue

In the preceding, I explored the role that synderesis plays in the development of the moral virtues. At this point, however, it is appropriate to return to the context in which the notion of synderesis is typically discussed, namely Aquinas's account of natural law. What conclusions, given the intimate connection between synderesis and virtue, can we draw about the relationship between virtue and the natural law? One important conclusion that the preceding analysis forces us to draw, I believe, is that natural law and virtue are far more intimately connected and far more interdependent than we are commonly inclined to think. For if the first speculative principles and synderesis are the primary explanatory principle behind the acquisition of the virtues, then it is clear that the moral virtues are directly dependent on man's natural habitual knowledge of the first principles of natural law. Without synderesis, it would be impossible for man to make any headway whatsoever in the acquisition of virtue. It is equally important, however, to note another implication of this view: without the virtues, man will find it very difficult to apply his natural habitual knowledge of the natural law in action. For Aquinas's own examples indicate that natural habitual knowledge of the first principles of natural law gives one little more guidance at the practical level than the first principles of speculative reason give one at the theoretical level. To truly be capable of rendering the principles of the natural law specific in action, one must cultivate the acquired virtues.

Might it be possible to make *some* headway in moral matters solely on the basis of one's natural habitual knowledge, without the acquired virtues? It certainly seems that some such headway can be made, but it equally important to understand that any such headway would be limited indeed. Important as the injunction to "do good and avoid evil" might be, it gives us very little genuine guidance. Indeed, it is hard to see how even the further precepts Aquinas offers, namely that one ought to do what is conducive to the preservation of one's life and of the species, could provide one with much real guidance in practical affairs. What the preceding analysis indicates, however, is that the natural law *should not* provide us with much guidance. The natural law gives man his fundamental orientation towards the good of reason, but genuine guidance in following its precepts can only come through the cultivation of the virtues. While this point is hardly an earth-shattering one, our previous analysis helps us to understand just how pivotal the role of synderesis is in the cultivation of the virtues, and thus helps us to understand why the moral virtues and the natural law are so intimately connected.

Notes

1. Pamela Hall's book certainly points in this direction, but since Hall's focus is mainly on Aquinas's explicit assertions about the dependence of prudence on synderesis, it is helpful to explore the broader connection between synderesis and moral virtue as a whole.
2. It is important, of course, to distinguish this from Aquinas's theory of the infused virtues, which are importantly different.
3. For an excellent discussion of this see Lear (1998).
4. Throughout the text the following abbreviations are employed: *ST*: *Summa theologiae*; *DV*: *Quaestiones disputatae de veritate*; *DQV*: *Quaestiones disputatae de virtutibus*.
5. It is worth noting that Aquinas uses slightly different wording in the *Summa* than he does in the parallel text in the *DQV*. In the *Summa*, Aquinas says that men naturally desire the good of reason, while in the *DQV* Aquinas says that men naturally desire the ultimate end.
6. Although Aquinas's discussion in *DV* is predominantly concerned with the acquisition of scientific knowledge, he gives a virtually identical explanation of the kind of potency man has for acquired virtue in *DQV*.

Bibliography

Aquinas, T. (1964). *Summa theologiae*. Blackfriars edition. Great Britain: Blackfriars.

Aquinas, T. (1994). *Quaestiones disputatae de veritate*, v. II. J. McGlynn (trans.). Indianapolis: Hackett.

Aquinas, T. (2005). *Quaestiones disputatae de virtutibus*, E.M. Atkins (trans.). Cambridge: Cambridge University Press.

Aristotle (1984). *The complete works of Aristotle*, J. Barnes (trans.). Princeton: Princeton University Press.

Hall, P. (1994). *Narrative and the natural law: an interpretation of Thomistic ethics*. Notre Dame: University of Notre Dame Press.

Lear, J. (1998). *Aristotle: the desire to understand*. Cambridge: Cambridge University Press.

Nelson, D.M. (1992). *The priority of prudence: virtue and natural law in Thomas Aquinas and the implications for modern ethics*. University Park: Pennsylvania State University Press.

Human Nature and Moral Goodness

Patrick Lee

How is human nature related to moral goodness? In what way is human nature normative? These are the questions I shall address from the perspective of a natural law theorist, or to be more specific, from the perspective of what is often called the new natural law theory, in the line of Germain Grisez, John Finnis, Joseph Boyle, and others. To approach the question, I shall first specify one way I think human nature is *not* morally normative. I shall then explain what I mean by *nature* when speaking of human nature, how I think free choice is related to human nature, how we reach basic practical insights and basic moral principles based on our nature, and I shall conclude by considering some theological worries about our natural knowledge of basic moral principles.

1 Human Nature is Not Itself the Moral Criterion

One idea of how human nature and morality are related is, in effect, that human nature just *is* the moral criterion. This idea can be expressed quite simply, in the popular arena, or developed in a sophisticated way (in, for example, the thought of Francisco Suarez and scholastic followers of him, who, nevertheless often mistakenly claimed they derived their view from St. Thomas Aquinas). In either case, the thought is that human nature is given, as a structure, and the biological and spiritual tendencies themselves which are parts of human nature, constitute the moral criterion in this sense: when one asks whether an action is right or wrong one simply compares the direction of the action to human nature or some part of human nature; if the action conforms to human nature—if basically the pattern of the action is the same as the pattern found in human nature—then the action is morally permissible; if the action does not conform to human nature then it is morally wrong. This, I think, is an oversimplified view of how human nature is normative or grounds moral principles.

P. Lee (✉)
Franciscan University of Steubenville, Steubenville, OH, USA
e-mail: plee@franciscan.edu

M.J. Cherry (ed.), *The Normativity of the Natural*, Philosophical Studies in Contemporary Culture 16, DOI 10.1007/978-90-481-2301-8_4,
© Springer Science+Business Media B.V. 2009

Take the example of the kind of argument this view has presented against lying. Proponents of this view might say that lying is wrong because it does not conform to the natural orientation or teleological direction in the faculty of speech—that the faculty of speech is naturally oriented to communicating what one thinks is true, but lying prevents that faculty or power from attaining its natural end and is therefore immoral. The central problem with this type of argument is that, given this conception of nature, it is not clear that we *are* morally required to keep within the guidelines, or limitations set for us, by human nature. If I use a basic power or faculty but direct it toward an end to which it is not naturally directed, why is that necessarily wrong? May it not just be ingenuity on my part, an end-run around the basic equipment I have been born with? Also, it is not clear that it is always wrong to act against every faculty or power natural to us—for example, it does not seem to be intrinsically wrong to induce vomiting, even, say, for experimental purposes, in order to discover the nature of digestion. (Let me add parenthetically, that I agree with many of the conclusions these arguments were supposed to support—I hold, for example, that lying, contraception, and extramarital sex are morally wrong—but I do not think this approach shows *why* these acts are wrong.)

Nor do I think that human nature is a norm the way in which some Aristotelians have argued (whether Aristotle himself held this position in every part of his ethics is a vexed question I will simply set aside here). Some Aristotelians have argued that human nature grounds morals by indicating to us the capacity that is distinctive of human beings, and, because distinctive, most excellent. And so, on this view, the morally ideal life is to develop to its utmost that part of us that is distinctive of us. I see several problems with this approach—though, I must apologize, that, for lack of space, I have stated it rather boldly and without, as they say, "nuances." In any case, the central problem with this approach is that it ignores the fact that it is only in the abstract that we share various capacities with other animals. True, both we and other animals are biologically alive and reproduce. But the biological life of a human being is fundamentally different in kind from the life of a dog or a cat. That is, in human beings there is not one half that is animal and the other half personal; rather, the life of a human being is through and through both a biological and personal life. This also is clear with reproduction: humans reproduce, but they reproduce in their own distinctive manner, and what is generated is not just a biological entity, but a whole human, personal being with both body and rational soul.

2 Human Nature

So much for how human nature is *not* the criterion. Now let us examine how human nature *is* a ground of practical and moral norms.

The nature of a being is its internal structure or set of tendencies that distinguishes it from other types of beings, coming from within, rather than imposed from outside. Some thinkers, notably existentialists, deconstructionists, or anti-realists, have denied that humans have a nature in the proper sense of the term. According to these thinkers, by our free choice or preferences, we construct what we most fundamentally are. Indeed, anti-realists of various types deny that there are real natures outside the human realm as well, holding that we construct or impose all meaning rather than discover any objective structures, or at least any important objective structures. It is clear, however, that when we begin to make something or to interact in any way with the external world, we must take account of pre-existing differences in those things we interact with. Prior to our projects and plans, different things have within them inherent tendencies to act and react in certain definite ways. Thus, we treat a tree differently than we treat a Rottweiler dog—we might kick a tree to vent our anger but will not likely kick a Rottweiler to vent our anger. The reason is obvious: within the Rottweiler dog there is an inherent tendency to act and react in a way fundamentally different than a tree. In other words, the dog has a different *nature* than the tree. Things in the world differ in that they are different types of *agents*—within each thing is a tendency, or an ordered set of tendencies, to act and react in certain ways. This inherent tendency or set of tendencies is the thing's nature. Aristotle spoke of nature as the intrinsic principle of motion and rest in a being, that is, the internal source of its tendencies to act and react in certain ways, or (put still another way) the internal source of the entity's basic potentialities.

Understood in this fashion it is clear that human beings *do* have a nature. Human beings have basic potentialities that are not acquired, not constructed, not imposed, but are inherent within them, basic potentialities that they have just by coming to be, that is, potentialities they have in virtue of what they are. And so human beings do have a nature. Human beings have the inherent potentialities to maintain their organic states, to grow, to reproduce, to perceive, to remember, to imagine, to have emotions, to understand, and to will. Not every human being can immediately exercise all of these potentialities, but every human being is oriented to actively developing himself or herself to the stage at which he or she will perform such actions.

3 Human Nature and Free Choice

Now to how human nature is related to free choice. I will bypass here most of the complicated disputes about responsibility and free choice, and presuppose that we do sometimes make free choices. Here I want to ask how our free choices are related to our nature. Both those who deny that we have a human nature, or that it has any role to play in morality, and those who hold simply that human nature itself is the moral norm, both view free choice and human nature as in tension or as opposed. Both groups view human nature as a limitation or restriction placed from the outside on our freedom or our choices. This is obvious for those who deny human nature or its significance, for they deny it in order to make room for freedom or choice. It also is true of those who view human nature as a pattern simply to be conformed with, for according to them it turns out that a morally good act is one that is within the limits of human nature and a morally bad act is one that strays outside those restrictions.

But this is a profoundly mistaken view of how free choice and nature are related. Free choice is not something that occurs as opposed to or as outside the framework set by one's nature. We choose among possibilities that are open to us, but what possibilities are open to us is set by our human nature. To express the point slightly differently: we choose *actions,* but actions are, in one way or another, actualizations of one or more of our basic potentialities. So, free choice occurs within the context of human nature—we do not choose to keep within human nature or to go outside it.

When I choose, I choose among possible courses of action. For example, I may choose to eat breakfast or to skip breakfast, I may choose to go to the gym to work out or to go home and study. Each of these possible courses of action is attractive, is of interest to me, because each, in one way or another, actualizes some basic potentiality in me. What makes a course of action attractive, what makes it desirable, and so in some basic sense, *good,* is that it is in some way fulfilling for me or those that I care about. A possible action is desirable or good (not yet morally good, but practically good) to the extent that it is, or at least seems to be, fulfilling (or is a means to an activity or condition that is, or seems to be, fulfilling).

So each choice is a choice to realize or actualize one's potentialities in some way or other. Since we are complex beings we have various basic potentialities. The actualizations of our basic potentialities are what natural-law theorists generally call basic human goods, or fundamental human goods. Such basic goods are the actualizations of the kind of being we are. So, freedom is not antithetical to nature; rather, our nature sets the possibilities within which we act. But our nature is not a limitation set upon our freedom or our desires either. Rather, the orientation to various types of fulfillment makes possible our choices and our active, free self-constitution in our choices.

I will return more fully to the issue of moral goodness shortly, but it is worth making a brief remark about it here at this abstract level. What could be the difference between morally good choices and morally bad choices, if whenever we choose we are choosing to pursue some basic good or other (or at least some fragment of a basic good)? Whenever we choose we are seeking to actualize one or more of our basic potentialities. Since, according to the natural law tradition goodness is the fullness of being possible for an entity, or the actualization of its potentialities, the point I just made can be rephrased as follows: whenever we choose we are seeking to realize some good or other. But if that is so, then how can we ever do what is morally wrong?

The answer is that there are two ways of seeking to realize one's potentialities, that is, two ways, of seeking fulfillment or a good. One can seek a good in such a way as to remain open to and respectful of all the goods not sought in this choice—and that is a morally good choice. Or one can seek a good in such a way as to turn away from, or diminish in oneself a respect for, some other good—either in oneself or in another, and that is a morally bad choice.

Thus, human nature is the ground or criterion for what is morally good without being a limitation on free choice. It is not as if morally good choices are those in line with nature and morally bad ones follow our desires and lead us outside the limits set by nature. Rather, all choices realize our potentialities (or develop our nature) in some way or other, but morally bad ones realize our potentialities in a way that diminishes our further openness to other basic goods. Nor is it that morally good choices are for the sake of good objects and morally bad ones are aimed at bad objects. Rather, all choices are for the sake of something to which we have some natural orientation, toward something good (or at least apparently good) but bad choices pursue a good in such a way as to suppress our appreciation of some other good (either in ourselves or in others).

4 Basic Practical Principles

The third point in our examination of how nature is related to morality is how we arrive at first practical principles—practical principles which are not yet, as I shall explain in a moment, fully moral principles. How, from these practical principles, we reach specifically moral principles, I shall examine in the next section.

Famously, David Hume held that any argument in which all of the premises are is-propositions, or descriptive propositions, and yet the conclusion is a moral ought proposition, commits a fallacy. Some natural law theorists have protested that this is *not* a fallacy, that propositions describing human nature together do imply moral propositions, and have held that the basic moral principles are deduced from propositions describing human nature. However, on this one point I agree with Hume (of course that is what, in part, puts me in the camp of the "new natural lawyers"). What Hume called a fallacy *is* a fallacy. For a practical proposition is a fundamentally different sort of proposition than a theoretical or speculative proposition—that is, an is-proposition. A theoretical proposition attempts to describe what is the case. In a theoretical proposition one is attempting to conform one's mind to what is. (I am presupposing here a realist position on theoretical propositions.) In a practical proposition, however, one is attempting to order acts of will, to put order in an act that has not yet come to be. A practical proposition is not a description but a directive or a *pre*scription. Now, in an argument the knowledge of the premises is the cause of the knowledge of the conclusion. And the effect cannot be greater than the cause. And so if all of the premises are telling us about what *is* the case, then, absent some implicit presupposition about what *ought to be* done, then those premises cannot by themselves generate knowledge about what ought to be done. (I do not say that some theoretical knowledge is not presupposed for practical and moral knowledge and arguments—to have ethical knowledge about killing and letting die, for example, one must know, at least in a general way, what life is, that we can do things to endanger or end life, and so on. But my contention is that the first principles of practical reason, and the first moral principles, are not established by arguments all of whose premises are theoretical, for example, theoretical propositions about human nature.) So, while moral norms are grounded in human nature, they are not deduced from propositions describing human nature. One need not first do philosophy of human nature before doing ethics. The first practical principles and the first moral principle are, in my view, self-evident and underived from any other propositions, even though they arise out of insight into the possibilities toward which we are oriented by our human nature. One could put it this way: they are *ontologically grounded* in human nature even though they are not *logically derived* from propositions about human nature. How, then, are they known, and how do they generate specifically *moral* propositions?

The key texts in the history of the natural law theory are, of course, found in St. Thomas Aquinas. According to Aquinas the first principles of practical reason are *per se nota*, that is, known through themselves, not deduced from any other propositions. And they are *per se nota*, or self-evident, because in each of them the predicate is contained within the intelligibility of the subject—so once one understands the subject one thereby grasps that the predicate belongs to it. An example in the theoretical order would be: if one knows what grass is, then one knows the truth of the proposition that grass is a plant; knowledge of the nature of the subject tells one that the predicate belongs to it. Something similar, but in a practical directive, occurs in the knowledge of the first practical principles. Aquinas has this to say about how these first practical principles are known:

> Since indeed good has the intelligibility (*ratio*) of end, and evil the intelligibility of its contrary, hence it is that all those things to which the human being has a natural inclination, reason naturally apprehends as good, and consequently, as to be pursued by action, and their contraries as evils and to be avoided (*Summa Theologiae,* I-II, q. 94, a.2).

In other words, by a practical insight one directly apprehends in the object of a natural inclination that this object is perfective or fulfilling and thus is good and to be pursued. This language of "to-be-pursued" (Aquinas uses the gerundive to express the prescriptive mode, as opposed to the indicative) indicates that the proposition is a directive, a prescriptive proposition, not a descriptive or theoretical proposition. I believe Aquinas's point here is correct. The basic practical propositions he is describing are propositions or directives toward the objects of our natural inclinations or our basic potentialities. Thus, we arrive at such propositions as: life is good and to be pursued, marriage is good and to be pursued, knowledge is good and to be pursued, friendship and society is a basic good and to be pursued, religion is good and to be pursued—these are the basic goods Aquinas lists (without claiming to provide a complete list). Following Grisez, Boyle and Finnis, I would add some others, such as play or skillful performance, aesthetic experience, and self-integration (see Grisez et al., 1988; Finnis et al., 1987, Chapters 9–11; Finnis 1983, 1998; Chappell, 1998). In any case, how do these insights with respect to the objects of our natural inclinations occur?

I think they occur as insights into experience of conditions or activities that are genuinely fulfilling. For, from the experience of these conditions or activities one comes to understand that, for example, health, knowledge, or harmonious relationships with other people, are of themselves fulfilling and thus worthy of being pursued. This insight is not an inference. Rather, if one grasps health, for example, as genuinely fulfilling, that is at the same to grasp it as to-be-pursued. An object's being genuinely fulfilling is the intelligibility under which one sees it as worthy of pursuit. The natural inclinations are, in a way, data for the practical insight. How this occurs may be illustrated, I think, in the following way. As a young child I experience being healthy as opposed to being sick or having my knees scraped up or having burnt fingers. I *enjoy*, or *take delight in*, being healthy, and I *dislike,* or *have an aversion to* being sick or wounded. That of course is not yet a practical insight; the enjoyment or aversion so far mentioned is on the level of emotion, or what Aquinas called sense appetite. But at some point as a child I go further and I come to *understand* that being healthy is a condition worth pursuing, and being sick or wounded is a condition I should take steps to avoid. What has occurred here is a practical insight, an act of intellect, an insight that being healthy is fulfilling and thus worthy of pursuit, protection, deliberation about, and so forth.[1] These first practical principles, it should be noted, are at the basis of *anyone's* practical deliberations. They are at the basis of the practical deliberations when we decide to do something morally good, but also at the basis of practical deliberations when we decide to do something morally bad. We cannot begin to think about what to do unless we understand some point, or points, to acting. Just as one will not begin to think about what road to take unless one sees some point in going somewhere, so one cannot begin to deliberate about what action to choose unless one first apprehends some goods worth pursuing. These first practical principles are just the knowledge that some activities or conditions are worth pursuing, promoting, or preserving.

5 From Practical Principles to the Basic Moral Principle

So, given the first *practical principles*, where does *morality* come in? The practical proposition that health is a good worthy of pursuit is not yet, or not yet formally, a *moral* proposition. The moral issue arises when, having seen and appreciated various basic human goods, such as health, friendship in a broad sense, understanding, and so on, one finds oneself in a situation where one could choose in two distinct manners. One could choose in a way that is fully in accord with all of these practical principles, or one could choose in a way that is in accord with at least one of them to a certain extent (else the option would not be attractive), but that is not fully in accord with all of them. And one's emotions have a role in making the unreasonable option attractive. Thus, *moral* normativity does not arise from an entirely new proposition or new contrast. The difference between morally good choices and morally bad choices is not that morally bad choices aim at something bad—badness is not a nature possessed by any thing but is a privation. Nor is the difference between the morally good and morally bad choices that morally bad ones are only in one's self interest and morally good ones are subjected to an implicit contract—though there are *some* morally bad choices which fit that description. Rather, because, as Augustine, Aquinas and many others pointed out, goodness is the fullness of being due a thing, the difference between morally good choices and morally bad ones is just that the good choices are fully rational, they have the fullness of being possible for a human choice, while the bad ones in one respect or another, are lacking in regard to the integral directiveness toward every aspect of human good that comes from the first practical principles in their totality and fullness.

So, morality and nature are closely interrelated, indeed, one can say that moral norms are grounded in human nature, even though they are not deduced from propositions describing human nature. However, to be clearer about the relationship, one perhaps should say that morality directs us to real human fulfillment (and I should add, not just individual fulfillment, but fulfillment in communion with other persons). Morality is based, not on human nature viewed as a static pattern or structure, but rather on genuine goods, the basic goods, to which we are directed by our nature. This clarification or qualification is important for at least three reasons. First, it makes it clear that natural law need not be conceived as a set of limitations or as primarily negative. On the contrary, when the emphasis is on the goods to which we are oriented by our nature, then the positive comes first: the basic moral norm is: energetically pursue and promote the basic human goods both in oneself and in others. The prohibitions, such as do not intentionally kill innocent persons, do not commit adultery, and so on, exclude choices and actions which are themselves negative. The second reason for the importance of this clarification (namely, morality is based on the goods we are naturally inclined to, not immediately on a nature)—is that it makes clear that we are not all called to have the exact same type of life; there is room for a great deal of diversity and ingenuity in living out the moral life. There are many basic goods, not just one, and there is perhaps an infinite number of ways of realizing these various basic goods. We are not all called to realize all the basic goods equally, or in the same order—we are called to fashion a life in which we energetically pursue some, and at all times to respect all of them, both in ourselves and in others. But natural law theory, correctly developed, does not prescribe that human beings attempt to be carbon copies of one another. And the third reason why this clarification is important is that it leaves room for supplement by faith and grace. The natural basic human good of religion invites us to inquire whether perhaps God might approach us to establish more than a natural harmony with him. The natural law directs us to *human* fulfillment, but it is, correctly understood, open to God's invitation to the supernatural, that is, something more than human fulfillment.

6 Human Nature and Divine Grace

So far I have not appealed to special revelation or religious faith. However, my claim that there are basic moral truths in principle knowable without appeal to special revelation, may cause worry, even consternation, among some Christians. Thus, H. Tristram Engelhardt, Jr., commenting on an article I wrote on abortion in the journal *Christian Bioethics*, complains about my "commitment to natural reason." He then says that, "This commitment to natural reason, characteristic of much of Roman Catholic bioethical analysis, tends to reduce moral theology and Christian philosophy to secular philosophy and to be distinguished by an accidental connection to the particularities of Christian commitments" (2004, p. 81). Thus, I would like to say something brief here, from a theological standpoint, in reply to the concern Engelhardt enunciates.

The goal of the Christian life is the completion of the Kingdom of God. As Our Lord says, "Seek first the kingdom of God and his righteousness" (Matt., 33; cf. Lk., 12:31). The Kingdom of God centrally involves a mysterious sharing in divine life. At Baptism we are made children of God, no longer just children of men. We have a "share of the divine nature", says St. Peter, and St. Paul speaks of "the Holy Spirit poured forth in our hearts." So, as Christians faithful to His invitation, we have in us, mysteriously, the divine life itself. This sharing in divine life can become more intimate and will become matured, or brought to fruition, in heaven, in the completed Kingdom of God (I believe Jesus was referring to this sharing in divine life, or grace, when to the Samaritan woman at the well He said, "And whoever drinks of the water I shall give him will never thirst; the water that I shall give him will become in him a spring of water welling up to eternal life" (Jn., 4:14).

But I believe also, first, that the eternal Kingdom begins now in mystery, that it is not just in the future, though it is completed only in the next world. For, Our Lord says that the Kingdom of heaven, or the Kingdom of God, is at hand, though it is like a mustard seed, and must grow. So we begin even now to share in, and with God's grace, to help build up, God's kingdom, or, with God's grace, prepare the materials for the eternal banquet.

Second, the Kingdom of God includes not only a personal communion with God, which is *super*natural (the *divine* nature in us), it also includes, as an essential component, *human* fulfillment. In the completed Kingdom there will be the mysterious, intimate personal communion with God, the seeing of God face to face, the beatific vision, but also complete *human* fulfillment. As a Catholic I heed the words of the Second Vatican Council, which said:

> For after we have obeyed the Lord, and in His Spirit nurtured on earth the values of human dignity, brotherhood and freedom, and indeed all the good fruits of our nature and enterprise, we will find them again, but freed of stain, burnished and transfigured, when Christ hands over to the Father: "a kingdom eternal and universal, a kingdom of truth and life, of holiness and grace, of justice, love and peace" (reference to the Liturgy of the Feast of Christ the King). On this earth that Kingdom is already present in mystery. When the Lord returns it will be brought into full flower.

Thus, the Christian life *does* include duties that transcend the fulfillment of human nature and cannot be known, even in principle, by natural reason unaided by the gift of faith. For example, we have a duty to accept God's Revelation handed on to us by Scripture and Church teaching, and this Revelation contains a proposal from God to enter into a personal communion with him, and to imitate Christ's life, and, with God's grace, to join ourselves to Christ's offering of himself on the cross for our sins, for it is by our being joined to that offering, that obedience unto death, that Sacrifice, that we are brought into that intimate personal communion with God, called grace or sharing in the divine nature. We are thus called to pray, to do penance for our sins, to participate in the Sacraments in which (according to Catholic belief) Christ's sacrificial offering on the cross is made present to us. Hence

Catholic Christians believe that Christians have *specific* duties not discoverable by the natural light of reason.

At the same time, if the Kingdom of God *does* include both human fulfillment as well as supernatural communion in divine life, it does not seem possible that the human part of what we are to work for as part of God's Kingdom would not be at least in principle discoverable by the natural light of reason. If it is part of *human* fulfillment, then it is proportionate to *human* nature, and thus it is hard to see why it could not be discovered, examined, reflected upon, and so forth, by the human intellect, even a human intellect that is darkened by sin and possibly in a human being who has not accepted, or has rejected, God's invitation to enter with him a personal communion. It might be very difficult for such a person to see some moral truths that are very difficult to follow, but I do not see how we could rule out in advance that possibility.

Still, we should add that the human intellect will certainly operate much more reliably, even as a human intellect, if it operates within the context or horizon of faith. There is no doubt that selfishness, pride, envy, lust and so on often bias the operations of the intellect. So, charity, humility, chastity, and, above all, faith, can help to heal that bias.

Note

1. Following this practical intellectual insight, there occurs what Aquinas refers to as *simplex voluntas*, that is, a simple act of will, in regard to an understood good—not a *choice*, which involves a willing to do something to bring about an instance of this good, but rather, simply a favorable inclination in the will toward the specific understood good. It seems to me that this *simplex voluntas* is part of the answer to the neo-Humeian argument that a reason for action must be motivational but that an understanding without an act of will is not motivational: the answer is to distinguish volitional acts—the practical principle is necessarily motivational with respect to a *simplex vountas*, but not necessarily motivational with respect to a choice.

Bibliography

Chappell, T.D.J. (1998). *Understanding human goods: A theory of ethics*. Edinburgh: Edinburgh University Press.

Finnis, J., Boyle, Jr., J.M., and Grisez, G. (1987). *Nuclear deterrence, morality and realism*. Oxford: Oxford University Press.

Finnis, J. (1983). *Fundamentals of ethics*. Washington, D.C.: Georgetown University Press.

Finnis, J. (1998). *Aquinas, moral, political, and legal theory*. Oxford: Oxford University Press.

Engelhardt, Jr., H.T. (2004). Moral knowledge: Some reflections on moral controversies, incompatible moral epistemologies, and the culture wars. *Christian Bioethics* 10: 79–103.

Grisez, G., Boyle, Jr. J.M. and Finnis, J. (1988). Practical principles, moral truth and ultimate ends. *American Journal of Jurisprudence* 33: 99–151.

Natural Law for Teaching Ethics: An Essential Tool and Not a Seamless Web

Jack Green Musselman

1 Why Learn and Teach Natural Law Today?

You really cannot judge that since it is all relative anyway.

Just about any ethics teacher in recent memory has heard this claim so frequently from a student, colleague or friend that he may well respond by just throwing his hands up in despair. That teacher has no doubt then wondered and worried that his entire job as a philosopher might well be pointless but still worth pursuing. It might seem pointless because despite vigorous and frequent attempts to show it is perhaps inconsistent for one's ethical challenger to maintain relativism—or at the very least that view becomes impractical as an ethical guiding star to live by—many people still insist that all moral standards are simply reducible (without loss) to the ethical codes that are strictly a function of our various cultural, historical and regional experiences. But pointing out the problems for that view is still worth pursuing, if only because one thing philosophers can still do is to argue how the relativist's main ethical claims and even his way of life ultimately fail to take advantage of the greater benefits of a more objectively grounded account of the right and the good.

As a result, many ethics teachers likely proceed by drawing from their toolboxes some standard arguments against ethical relativism.[1] Well known to any ethics teacher, these arguments include pointing out how a relativist would have a hard, even impossible, time of defending his own moral beliefs as much better than another's, or even end up grounding those beliefs in a circular or subjectivist fashion by replacing the latter part of "My moral beliefs are correct" with "My moral beliefs are simply correct for me here and now in my life in this cultural moment." While not obviously false such accounts from most students who typically do not really think (deep down when pressed on the matter) that their moral commitments are really so ephemeral or so weakly constitutive of their agency would also take much

J. Green Musselman (✉)
Department of Philosophy and Center for Ethics and Leadership, St. Edward's University, Austin, TX, USA
e-mail: jackgm@stedwards.edu

M.J. Cherry (ed.), *The Normativity of the Natural*, Philosophical Studies in Contemporary Culture 16, DOI 10.1007/978-90-481-2301-8_5,
© Springer Science+Business Media B.V. 2009

more of their time to defend than the few minutes of hand-waving that the instructor might usually encounter in his class.[2]

And while convincing for many of us, most teachers of ethics can attest that such philosophical advances often do not make much headway with everyone from undergraduates to the neighbors next door. There are likely several reasons for such pedagogical shortcomings, including our all too human psychological commitments to ways of life learned over many years and a postmodern culture which makes discussing (much less reaching) coherent and rational ethical consensus very difficult today.

As to the former, one suspects the reason that many people—perhaps especially but not limited to young men and women—do not immediately reject ethical relativism upon a showing of its conceptual or practical challenges is that commitment to such an ethical framework is in the first place less philosophical than psychological. After many years of at least *thinking* one had not adopted such a thoroughly relativistic code of values, upon learning that this very code was untenable a young student of ethics would likely think and *feel* he had yet to make a difficult intellectual and *emotional* transition to a more solidly grounded system of ethics. That is, the cognitive change from any ethical relativism to a more objectively grounded set of values would likely also require an *affective* shift that amounts to changing identities, too, so that to do this successfully the young student of ethics would also have to *mourn* the loss of the old relativistic self and construct a new self based on a more solid ethical footing.[3] If this is correct, it is no wonder that showing the philosophical and practical challenges of believing in (and living by) ethical relativism does not lead young (and older) students of ethics to reconstruct their very selves as quickly as possible as they search for a more objective code of ethics.

If such ethical agency construction is hard for individuals to manage, reaching a social consensus on what truly counts as a theory of the right and the good would be even more difficult. Even if young (and old) students could face these challenges and reach an ethical consensus about the good life, that would of course no more (by itself) make such an account objectively good than simply getting people to agree that $2+2=4$ would make that claim (without more) mathematically true. Nonetheless, reaching a social consensus on an (otherwise) objectively grounded system of ethics would make it more likely that our society would secure for more of its members a life truly marked by fair treatment and the pursuit of just ends—no small feat for college students today and in the future who may well be called to assess everything from cloning humans in the lab to steroid use in competitive sports.

Even if social consensus about ethical goods did begin when young students struggle to alter their very identities in search of new and more tenable ethical beliefs, our postmodern culture does not make one optimistic that we will reach this ethically and socially valuable consensus on a coherent account for an objective system of ethics. For starters fewer young students seem to attend college today to develop a meaningful philosophy of life but rather seem more likely to matriculate to develop skills they can parlay into a job once they obtain a degree.[4] Even without thinking that today's college should produce only thinkers and not doers, it is not much of a stretch to claim that a postmodern culture that makes virtual chat look more attractive than face-to-face dialogue and rewards college students who use amphetamines and Ritalin just to obtain higher grades[5] would also be a culture that makes authentic and reasonable interpersonal consensus on objective ethical norms difficult if not impossible.

Quite apart from the way postmodern culture frames for students how their choices make and unmake the world, one should also note that the current culture makes ethical discussion at least as difficult as when Alasdair MacIntrye lamented nearly twenty years ago that our then-current "moral disorder" took the ethical traditions of Plato and Hume, Kant and Mill out of their historical context of shared traditions that made ethical disagreement more likely to resolve (1984, p. 11). That is, if the lack of shared traditions and practices constitutes and characterizes a moral schizophrenia of our current cultural era beyond the repair of many traditional ethical theories, so too one wonders if the postmodern culture also makes impossible (at least MacIntyre's) modern virtue theory and its rewards of excellence internal to the exercise of virtuous practices. That is, could such modern virtue theory all by itself help young students reach a consensus about truly objective moral norms if they lack the knowledge of the historical traditions of virtuous practices that the theory itself requires?

Perhaps before begging the question about virtue theory a more robust conception of that theory ought to be briefly sketched here. For such purposes, such a brief conception might benefit from the trenchant remarks by Philippa Foot on the virtues (1998). As Foot begins her essay by making some "fragmentary" remarks about moral virtues that she takes to be fairly uncontroversial, that will be a good place to start.

First, Foot argues that we "do not get on well without" the moral virtues such as courage and wisdom and "communities where justice and charity are lacking are apt to be wretched places to live" (p. 164). Second, for Foot such virtue "belongs to the will" (p. 165) but in the "widest sense, to cover what is wished for as well as what is sought" (p. 166). This "widest sense" means that the wisest human knows not just the means to the good ends but he also knows and judges the true value of the ends to pursue; not pursing, say, money and fame to the detriment of family and friends (p. 168). This "widest sense" also implies that, unlike an account of arts or skills in which a deliberate error (e.g., a grammar teacher's intentional spelling mistake for pedagogical purposes) does not count against the art or skill, a deliberately vicious act engages the will and so does not excuse the immoral agent from blame (pp. 168–169). In short, when exercised the moral virtues are good for the actor and his community, and the virtues he exemplifies belong to the will in the sense that they indicate how he achieves his ends and the value of the goals he seeks in ways that make him blameworthy for intentionally doing the wrong thing.

Before moving to Foot's next few claims about the virtues it would help here to adopt MacIntyre's definition of virtue:

> *A virtue is an acquired human quality the possession and the exercise of which tends to enable us to achieve those goods* which are internal to practices and the lack of which effectively prevents us from achieving any such goods. . . .we have to accept as necessary components of any practice with internal goods and standards of excellence the virtues of justice, courage, and honesty (1984, p. 191).

Like most teachers of ethics and indeed for MacIntyre too, it helps to spell out this definition by analogy to chess, basketball or piano to suggest that there are traditions of excellence in which we must train to learn the practice and gain the worthwhile internal goods. So while we might at first take piano lessons or play basketball games (or do the right thing) to impress family and friends, eventually when well-played we earn the intrinsic reward of these practices much like the moral actor earns the flourishing of a life well-lived by subsuming her early and late practice of the virtues to the traditions in which they are inscribed.

With MacIntrye's definition in place, an account including Foot's additional commentary on the moral virtues might emphasize several claims intended to flesh out how the virtues work when confronted by challenges as well.

First, Foot claims that the virtues are "*corrective*, each one" helping us resist some "temptation to be resisted or deficiency of motivation to be made good" (p. 169). Thus, for Foot industriousness is a virtue because we are tempted by idleness, humility is a virtue because we can too highly praise our own talents, and hope is a virtue because we too easily are led astray by despair (p. 170)[6]—as the suicide victim exhibits even when he does not leave behind family harmed by his absence (p. 174). Of course, some situations might present different temptations and obstacles, and some people may have more sophisticated natural abilities to resist or overcome all such things. However, such cases and situations might only make us say that some virtues are present or incomplete and not that they are completely inadequate or lacking (p. 171).

For example, Foot acknowledges that a case of stealing that is tempting for a rich and poor man might make us judge this situation and the characters of different actors differently, especially if the obstacle were great to overcome but not quite in the same way for both. Likewise, we might think the charitable person who gives up too much, and too easily, to a rival is more virtuous than one who does not because charity, as a virtue of attachment and action, is measured in part by the obstacles we overcome by having sympathy for another and then acting accordingly (p. 172).

Second, if an "action with 'positive moral worth' " or a "positively good action" is in "accordance with virtue" and is "one for which a virtue was required," what happens when, say, "displaying one virtue" in the face of temptations, such as "courage or temperance or industry" advances, say, "in an act of folly or villainy" (p. 174)? Or can a bad action display the virtues (p. 175)?[7] For Foot, we might think the villainous murderer's homicide "took courage" but this does not make his "act of villainy" courageous itself just because he was tempted by danger not to act (p. 175). By comparison Foot suggests that as a poison is a property of physical things but those things are not always acting as poisons (perhaps when used for a good end), courage is a property of human acts but those acts are not always courageous (perhaps when advancing a bad end) (p. 176). Also, by way of defense, Foot suggests that not every person who has a virtue has a feature that is a virtue in him, as say a person who is industrious might well be over-industrious to the point that he denies the other pleasures in his life. Thus, here we might say "the virtue did not operate as a virtue in this man" (p. 176).

There are plenty of other things that one could say about this account of virtue theory, for after all such a short sketch is certainly incomplete. However, considering just the virtue of honesty suggests, for starters, how unclear are the some of the historical contexts for determining the so-called objective boundaries of this virtue in ways that might help students reach some socially useful consensus.

We might begin by arguing that honesty is certainly a learned practice that we first try on for size as children. That is, with some prompting we learn from our parents that it is important to report honestly when we have been unkind to a friend in the sandbox and also honestly to indicate when after meals we have failed to brush our teeth. On this account of virtue theory, we may report about such things truthfully at first to gain the external reward of our parents' approval but later do so to earn the internal reward of practicing honesty in our daily lives. While learning the historically-embedded practices of honesty within our families, Foot might add, honesty as a virtue is certainly a corrective when telling the truth successfully combats and trumps the temptation to lie about not hitting our little friend when such a lie could very well let us continue to aggressively play in the sandbox as we see fit. In addition, a deliberate lie just to advance such selfish ends might be considered more improper than just, say, forgetting we had not really brushed our teeth, because the intentional lie engaged the will and, as Foot reminds us, hurts the members of society—or our family—because they get on less well for not being easily and effectively able to pursue the intrinsic rewards of family life as honest and kind people who are bound by love and trust.

But as convincing as this short sketch might be, there are likely many historically embedded traditions of truth-telling in which the virtues are inscribed and it is unclear which of these traditions most objectively ground which virtues' rewards internal to particular practices with their unique standards of excellence. For example, if most twenty-first century Americans agreed that the internal rewards from toddlers practicing the virtue of honesty requires truth-telling in the fairly mundane cases of being mean in the sandbox and about brushing their teeth, we might still have very different historically-informed accounts of the temptations (if any at all) such honesty overcame. That is, we might have different historical accounts based on our own experiences in childhoods about how easy or hard such temptations are to overcome, and certainly the same historical reflection might suggest that some children more quickly learn to trump the temptation to lie due to no willful practice of their own but rather due to their biological disposition, moral luck, exposure to narratives in books and movies about the effects of lying and so forth.

Likewise, when practiced in our different historically embedded families the virtue of honesty might be variously characterized by different internal rewards in situations where temptations to lie are present and even strong. For example, consider the different plausible characterizations of virtuous practice for children who learn to avoid the public faux pas of speaking fully and frankly when they instead side-step Aunt Mame's question about exactly what they thought of her (privately) detested Christmas fruitcake.[8] That is, here it is unclear if honesty is a practice with its own distinct internal reward if we combat and overcome the temptation to be brutally honest and instead tell a little white lie (I cannot even begin to describe this year's fruitcake) to better honor family loyalty and love, or if honesty is a practice with a distinct internal reward if we combat the temptation to tell the little white lie and instead offer a more honest remark (It is just not as good as the really tasty chocolate cake Aunt Kate made last year) more effectively to earn the internal reward earned by engaging in more obviously bold practices of truth-telling. That is, which historical practices should inform our accounts of virtues when they conflict? Should we say that the virtue of honesty is present in both cases but lacking or incomplete in one of them? Is there a second-order trans-historical account of virtue theory that would tell us which set of practices is more likely to earn the most appropriate internal reward? Which account of virtue theory should be adopted so as to get on better as a family, much less a city or country? Far from eliminating the problems mentioned above, historicizing the practices of the virtues to achieve the relevant excellence seems to re-introduce the problems of cultural relativism that led us to virtue theory in the first place. If that is true, the problems are only compounded if there are different culturally and historically determined practices of virtues on which to draw as opposed to a set of trans-cultural traditions that are most objectively true for providing the most robust internal rewards for the right virtuous practices.

Any ethics instructor today will recognize this challenge for virtue theory, and similar challenges to Kantian and Utilitarian and relativist approaches as well. This might then lead teacher and student alike to probe deeper about the nature of the *human beings* who are seeking these ethical answers. That is, students and teachers alike might well then ask about the *nature* of the humans who are philosophically and psychologically dissatisfied with reducing or grounding their ethical claims to or on such cultural assertions. They might ask about the human beings who can gain the internal rewards of honesty by side-stepping a more frank reply to Aunt Mame about her fruitcake and perhaps gain similar internal rewards by telling the truth to their parents and friends about being mean in the sandbox and workplace. The teacher and student of ethics might inquire about the nature of human beings who should allegedly seek or obtain only pleasure as a moral good as Mill would require, and about the nature of human beings who should when constructing ethical replies minimize all inclination and appetites and instead appeal to their rational nature alone as they attempt to universalize maxims as Kant would require. That is, what do such ethical theories do to and for human beings and to those with whom we live?

It is when ethics teachers and students alike ask such questions about the nature and value of the pursuits of human beings that they both are ready for and need to examine the claims of natural law theorists. Patrick Lee's and Christopher Tollefsen's essays in this volume suggest how two natural law approaches can advance the theoretical and applied dimensions of that very same ethical dialogue. In the pages to follow, I will outline the main arguments in both essays as well as make some brief suggestions about what teachers and students of ethics might both learn from reading and studying the natural law perspectives raised and defended in their two essays.

2 Lee and Tollefsen on Natural Law

These two thoughtful and cogent contributors advance the ethical discussion of natural law in different but complimentary ways. Lee's essay "Human Nature and Moral Goodness" provides a theoretical account of how "human nature is related to moral goodness" and the way human nature is normative (Lee, p. 45). Tollefsen in "Human Nature and Its Limits" focuses on developing a theoretical approach to natural law to assess the moral status of medical technological enhancements promised by cloning and brain-enhancing drugs and implants (Tollefsen, p. 17).

Lee begins his essay with the qualification that he is not adopting the "oversimplified view" that "asks whether an action is right or wrong" simply by determining if it "conforms" with human nature or is a "pattern found in human nature" (Lee, p. 45). For example, on this view if speech is "naturally oriented" to communicating the truth then lying thwarts that natural end of speech. Lee quite rightly rejects the view that it is always "wrong to act against every faculty or power natural to us" by himself pointing out that it would not necessarily be "intrinsically wrong" to take an emetic in order to learn more about how digestion works (Lee, p. 46). Such qualifications will be welcome to teachers of ethics who might only encounter in the classroom a straw person account of natural law painting it as simply endorsing whatever is commonly found in nature, is consistent with a single solitary function of a human's particular biological activity, or simply equates natural with good itself.[9]

Instead, Lee intends to ground *nature* for practical and moral norms by claiming its inherently pre-existent "internal structure," "potentialities" or "ordered. . .set of tendencies" help differentiate the nature of a being (from the inside-out) from the nature of other kinds of beings by influencing how each acts and reacts on its own (Lee, p. 47). So just as dogs and trees act a certain way because of these intrinsic structures or tendencies or potentialities, humans are "oriented" to reproduce, recall, understand and will. Of course since humans also make free choices we can choose to actualize a potentiality or not actualize it, and by actualizing it we achieve one of our "basic" or "fundamental human goods" (Lee, p. 48).

Lee next proceeds to address the Humean naturalistic fallacy of not deriving a moral claim from an empirical one by arguing that his own "first practical principles"—ones that are not yet completely natural law moral norms—while being "self-evident and underived from any other propositions" are "grounded in human nature" but are "not deduced from propositions describing human nature."[10] But, Lee asks, how are these "first practical principles" then "known, and how do they generate specifically"—if only by this point yet incompletely—"*moral* propositions?" (Lee, p. 50). The answer, following Aquinas, is that by "practical insight one directly apprehends in the object of a natural inclination"—or of our "basic potentialities"—"that this object is perfective or fulfilling and thus is good to be pursued" (Lee, p. 50). So, we conclude that "life is good and to be pursued," and likewise knowledge, friendship, marriage, play, aesthetic experience and self-integration are good and to be pursued.

Given this rather long list of self-evident (but as yet incompletely moral) potentialities to be actualized, Lee is right to ask how we so capably achieve such "insights" about the "objects" of all our "natural inclinations." And perhaps because there is really nowhere else to turn, Lee suggests we achieve such insights by and into and immediately (but not by inference) from "experience of conditions or activities that are genuinely fulfilling" (Lee, p. 50). For example, if we experience being sick and being healthy we "*enjoy,* or *take delight in*" being healthy and then "come to *understand* that being healthy is a condition worth pursuing" (Lee p. 51; emphasis in original). We follow these steps before we make any decision—moral or not—to see that there is some good to be achieved or realized, or some *point* to acting, which is achieved as a "practical insight" by "an act of intellect" (Lee, p. 51).

Thus far Lee admits his first "*practical principles*" do not yet fully constitute moral principles. What, then, makes the difference here is *first* that the "basic human goods, such as health, friendship in a broad sense, understanding, and so on" require choices that either fully realize all such goods or some but not all, *and* good moral choices are "fully rational" and have the "fullness of being possible for a human choice" (Lee, p. 51). Or, as Lee claims, a moral way of realizing human goods leaves one "open to and respectful of *all* the goods not sought in this choice" (Lee, p. 48; emphasis mine). By contrast, bad moral choices lack "the integral directiveness toward *every* aspect of human good that comes from the first practical principles in their totality and fullness" (Lee, p. 52; emphasis mine), no doubt when bad moral choices actualize a potentiality in "such a way as to turn away from, or diminish in oneself respect for, some other good (either in oneself or in another)" (Lee, p. 48). This totality and fullness require not just a person's solitary fulfillment but rather "fulfillment in communion with other persons" (Lee, p. 52) as we achieve the full set of complex "basic goods, to which we are directed by our nature" (Lee, p. 52). Lee then adds that such a natural law account of morality need not claim we are "called to have the exact same type of life," thereby leaving "room for a great deal of diversity and ingenuity in living out the moral life" —including leaving "room for supplement by faith and grace" (Lee, p. 52).

How, then, could a teacher and student of ethics profit from such a defense of natural law ethics? Lee suggests one relevant example by outlining the goods of friendship. Certainly a virtue theorist, or Kantian or Utilitarian, would not be so utterly incapable of accounting for the goods of friendship. However, as any ethics teacher can attest students sometimes find these accounts wanting—and for good reasons. For example, on one account of virtue theory a friendship might best be created with the Other by earning the intrinsic rewards inherent to practicing the virtues of honesty and justice with him, but of course a friendship might well require loyalty too, thereby making it very difficult to act out of tough love to tell the Other some hard truths he would (otherwise) be better off confronting. That is, as far back as Socrates questioning Euthyphro on his way to prosecute his father for murder we have reason to wonder how the goods of friendship inside *or* outside a family could include the virtues of loyalty that are sometimes so terribly at odds with the virtues of honesty or justice. Likewise, the goods of friendship seem incompletely captured by parsing them as mere attempts to maximize pleasure or universalize maxims, since those goods seem like they should be characterized in more robust ways than that to honor their place in the lives of human beings.

By contrast, students and teachers alike might wish to adopt Lee's natural law framework and suggest that friendships, perhaps a mutual association based on similar interests and values and commitments to the Other's good, are pre-existent "potentialities" or "ordered...set of tendencies" that by free choice we actualize to realize personal *and* social goods. Friendship is not good primarily because it produces pleasure or can be universalized or earns its practitioners intrinsic rewards, but rather because it fulfills a very individual and communal good that we realize through the "experience of conditions or activities that are genuinely fulfilling." Of course, not all friendships are immediately morally good ones, as when lying for a friend under oath in a court of law one acts toward the community in "such a way as to turn away from, or diminish in oneself respect for, some other good (either in oneself or in another)." Instead, for a friendship to more perfectly realize a moral human good one should not necessarily lie, for that act fails to honor friendship's "fulfillment in communion with other persons" who are not themselves one's immediate friends—viz., the loved ones and community members in the courtroom who are not honored by the lie itself but still stand in relation to self and Other as members of the broader human community.

In what other ways might students and teachers alike wish to adopt Lee's natural law framework? Certainly friendship, marriage, aesthetic experience, play and religion might well be amenable to such a natural law analysis, for by making choices required for achieving the goods of each we might need to remain open to all others and especially those not achieved if, in fact, such morally good choices require an "integral directiveness" (Lee, p. 52) toward all of the goods inherent to each experience.

However, could Lee's analysis lend itself to defending for a natural law theorist the otherwise counter-intuitive moral endorsement of consensual non-marital sexual acts between people of the same or different sex?

To begin this analysis Lee would have to concede that from the mere fact that such non-marital sexual acts occur as "a pattern" found in human nature does not, by itself, tell us anything about the moral status of such acts. Instead, these non-marital sexual acts must be assessed by first determining if there were an inherently pre-existent internal structure, "potentialities" or "ordered...set of tendencies" to perform them. Since people do, in fact, engage in such non-marital sexual acts, even if they are not considered mere patterns they might well be called a potential or even sets of tendencies we realize in some way or another. Of course, even this is not enough under Lee's natural law analysis to label as moral the self-actualized tendencies or potential to engage in non-martial sexual acts. Several more steps are needed.

First, are non-marital sexual acts "perfective" or "fulfilling" of some object or good which we know from experience, and then we come to understand by act of intellect that we now know this object is worth pursuing? Certainly I experience health as good and then, by practical insight, can know it is a goal worth pursuing because I do not enjoy and delight in being sick. But something very similar might be said of, say, heterosexual New York bachelor John's non-marital sexual acts for demonstrating, perfecting and even fulfilling the goods called intimacy and love for the Other adult person, and John may also thereby know he does not enjoy restricting his achievement of these goods with this other person by abstinence. That is, even outside the bonds of marriage one might argue that our experience of such non-martial sexual acts helps us, after the fact when we use the intellect, to understand that we can realize these goods through direct experience.

Second, are these non-marital sexual acts with the Other adult person choices that fail to leave us open to, or even realize, most or all the other natural goods for the self or others by having us turn away from the fulfillment of these other goods? Does realizing the so-called goods of non-marital sexual acts diminish an agent's respect for these other goods by perhaps focusing an agent on his self-fulfillment at the expense of social communion? Without begging the question by assuming all such goods merely come packaged in a seamless web experienced either altogether or not at all, it is not clear at all that New York bachelor John would not be capable of experiencing any, most or the fullest range of goods inherent to knowledge, friendship, religion and aesthetic experience—or that he would diminish some other single or faithfully married woman's fulfillment of the same goods—merely because he has chosen an allegedly less than complete sexual act to perform outside of marriage.

For example, it is not clear why the unmarried sexual agent John could not experience—with or without the faithfully married Mary living in the New York house next door—the full range of divine aesthetic beauty and perfection offered by a majestic view of the Grand Canyon in the American West or something similar offered by sitting in the preternatural blue glow of the stained glass windows of Marc Chagall's chapel in Nice, France. That is, why would John's non-marital sexual acts with the Other adult make him less capable of the full range of aesthetic experience provided by the Grand Canyon's unequaled vast natural expanse or by the ethereal blue of Chagall's chapel windows? It would seem that John can just as well still be a close friend of Mary's, enjoy the communal experience with her of being transformed by God's nature or Chagall's art, and even discuss knowledgably the rivers that formed the former and the craft techniques the latter in ways that *enhance* these very aesthetic experiences. So even if it were true that John's non-marital sexual acts are the less than perfectly fulfilling sexual acts within a marriage, he does not seem, without more, to be less open, or make others less open, to the other natural goods available to him or others in this community.

Third, it is certainly still possible that John's non-marital sex acts cost him the more natural "self-integration" gained by, say, his faithfully married New York friend Francine who engages in sexual relations with her husband Frank. So while Francine's marital acts "realize the unity of marriage, which includes the coming to be of children," John may do damage to the "unity, the integrity of the marriage" when he treats his body, a "part of the personal reality of the human being" as a "mere sub-personal instrument" to be used and disposed of to satisfy the subjective wants of the conscious and desiring part of the "self." On this view, John's "psychosomatic integrity" is this basic intrinsic good that "is disrupted" by his non-martial sexual acts that lack the "common good of marriage as" their "central specifying point." So when John engages in non-marital sex "purely for pleasure, or as a means of inducing feelings of emotional closeness" he experiences an "existential separation of the body and the conscious and desiring part of the self" that "literally" disintegrates or "takes the person apart, disrupting the good of acting as the dynamically unified being one truly is."[11]

But even assuming John (consciously or not) experiences this lack of self-integration in non-marital sex it is unclear how and why his entire agency or personhood is so broken into pieces that he then fails to achieve the "the integral directiveness toward *every* aspect of human good"—from aesthetic to knowledge to friendship—that comes from his moral failure to achieve the perfective goods constituted by sex within marriage. What is missing from Lee's otherwise rich essay, I suspect, is a more robust account of how a failure of one natural moral good more immediately and directly causes one to fail to experience or achieve (as a solitary agent or in social union) the other moral goods as well.

One kind of more robust account could even be constructed for all these New Yorkers, from John and Mary to Francine and Frank. Suppose John and Mary engage in a clandestine extra-marital sexual affair. On this scenario their individual friendships with her husband, Henry, could very well suffer if John and Mary were as a result of their affair not open to the goods secured by honestly and openly sharing with Henry their deepest hopes and fears. (And the goods of friendship that are achieved on this scenario are hardly as full as they might be given they are premised upon intentional deception.) Moreover, even if John and Mary at first secured from Henry his full and free consent for their extra-marital affair, on this scenario the jealousy and anger this might still engender could easily prevent Henry from achieving with them the goods of friendship or keep him from achieving the good experienced from successfully engaging in communal worship at Sunday religious services with (or without) Mary. Furthermore, if Henry and John were Roman Catholic their consent to this unorthodox marital arrangement would constitute a sinful choice that would close their hearts to God and would and should, if they pursued this arrangement unrepentantly, prevent them from achieving the state of grace necessary for engaging in Holy Communion so as to know and receive God during church services.

However, even this brief, more robust account does not, strictly speaking, philosophically demonstrate how a failure to actualize one moral good prevents individuals in most or every case from achieving the other moral goods as well. The New York scenario in which Henry, John and Mary consent to the sexual affair might well *not* lead to anger and jealousy depending on the psychological states of all three people as their unorthodox extra-marital relationships are pursued—even at the very same time as they individually and collectively attempt to gain knowledge and aesthetic goods. This would seem to be an almost entirely empirical matter, for it may be that their friendship is *not* harmed by the unusual sexual liason any more than their ability to enjoy together the full aesthetic potential of the Grand Canyon and Chagall's chapel. That is, it is not clear why Henry can not (and not in bad faith) successfully achieve the good of friendship with Mary and John despite the unusual arrangements, or why he might not still share his deepest hopes and fears and even enjoy Chagall's sublime chapel blue or *Macbeth*'s tragedy despite the real risks run by this consensual extra-marital sexual affair.

Moreover, it might well be true that New Yorkers Henry and John cannot properly as Roman Catholics open their hearts to God and receive the Eucharist when their full and free choice to consent to the extra marital sexual affair is not confessed and they do not repent of this willful fall from grace. But even if this is true as a matter of Roman Catholic theology it is not necessarily true of all kinds of Christian religious practice, to say nothing of other forms of religious worship. So even if it is true that Henry and John can sustain their friendship and experience aesthetic joy while consenting to this unorthodox extra-marital sexual practice, it is not clear why the goods of religion *qua* religion are for them foreclosed. It might be a more robust account of Catholic theology and religious practice would make clear why this unorthodox extra-marital affair prevents them from securing the goods of religion as such, *while simultaneously* failing to achieve the moral goods of true friendship *and* rich aesthetic experience—for individuals alone and people communally in groups—but that would take a longer and more developed argument than presented here.

Another interesting and valuable account for teachers and students alike is Tollefsen's treatment of how natural law ethics might address enhancements in modern-day technology and medicine. That is, if chemicals or silicon can be used to increase human memory or intelligence, would such modern-day experiments immorally alter our human nature?

To begin his argument Tollefsen offers an account of human nature or identity on which we are first biological humans (and so directly experience the world as such) and then also persons (who have capacities of reason and choice) (Tollefsen, pp. 18 and 19). This practical identity develops over time and is "self-constituting" in the sense that what we are we become by choosing what we can be. Like Lee, who emphasizes both individual and social dimensions of the human good, Tollefsen claims that the ground of how we act and can create our very selves is "found in the well-being and fulfillment of human persons and communities they form." While this human fulfillment and good are diverse, Tollefsen argues there are "many *irreducible* dimensions of human well-being and flourishing" and so a "variety of reasons" or "human goods" that thereby "establish the more than given essence of human nature" (Tollefsen, p. 20; emphasis mine).

Continuing to follow in the natural law tradition, Tollefsen calls the "most fundamental dimensions of our well-being *basic goods*" because they advance human flourishing in ways that give us "reasons…to act that do not need grounding in some further reason" and so "perfect and fulfill" all of us "as human beings" and as persons (Tollefsen, p. 21; emphasis in original). Among these basic goods Tollefsen, like Lee, includes friendship and health and art that serve individual as well as social purposes, and knowledge, the last giving "meaning" to how we act and thus not requiring additional justification (Tollefsen, p. 21). Since these goods are part of a complex human nature, they also require we harmonize our wills with others and with one's own source of "transcendent meaning," "practical reason" and "emotions and dispositions" (Tollefsen, p. 21). Much like Lee, Tollefsen here adds that our "practical nature" specifies and limits the "basic goods that constitute the horizon of our possible flourishing" to a "finite, relatively small, and permanent" set of goods made possible "by our own action," or by our autonomous "deliberation, choice and commitment" (Tollefsen, pp. 21–22).

The remainder of Tollefsen's article primarily examines how enhancements in technology and medicine might well immorally threaten our fulfillment as human beings and persons, our ability to pursue these basic goods and to achieve them by our "creative activity," and even disparage "the particularity of our animal nature" to such a degree that we cannot achieve these basic goods in the first place (Tollefsen, p. 23).

First, scientific breakthroughs in reproductive cloning—once the provenance of science fiction but maybe no longer so far away—remove sexual reproduction "from any necessarily interpersonal context at all." But since "sex and marriage are other-related, interpersonal realities" and our flourishing requires "relation to others, particularly in goods as friendship, marriage and the family," technologies that make reproduction possible without sex and marriage are "reductive" of our possibilities for well-being by taking the necessary interpersonal Other out of the picture. Likewise cloning is too "self-focused" for making us "autonomous procreators," distancing us essentially from the necessary interpersonal Other needed for our well-being and engaging us in a "lonely and solipsistic business" that takes us away from "others with whom we would have had the potential for interpersonal communion" (Tollefsen, p. 23). Likewise, the basic goods of "loving, human relationships" realized by reproductive sex would be lost by cloning's "designers and manufacturers" who would "alienate themselves from their children" (Tollefsen, p. 23).

In what ways might students and teachers alike wish to adopt Tollefsen's natural law framework? Tollefsen here offers ways that health, friendship, marriage, art, knowledge and religion (as one main source of transcendent meaning) might well be profitably grounded by a natural law analysis that simply can not be captured quite as well by Kantian, Utilitarian or virtue theory accounts of ethics. However, as with Lee's analysis one wonders if Tollefsen's own account lends itself to defending for a natural law theorist the otherwise counter-intuitive moral endorsement of consensual non-marital sexual acts between people of the same or different sex.

For example, consider a woman named Sally and her good friend Jack. Suppose Sally lives next door to Jack in New Jersey, though now imagine that Jack has a state-sanctioned civil union with his long-term life partner Alex. Both Jack and Alex might agree they are biological beings who are persons capable of choice for self-constituting as individuals and who find fulfillment by living in larger communities beyond their front door. When discussing the basic goods and their irreducibly human characteristics Jack and Alex identify friendship and art, religion and knowledge as not requiring further justification, for they advance self-fulfillment and essential social goods interdependently when Jack and Alex decide to join their friend Sally as they stand in awe before Chagall's chapel and join her as their church builds a house over the weekend with Habitat for Humanity. All three might also realize that there are a finite possible number of goods they can achieve together by harmonizing their wills autonomously to achieve the human goods associated with friendship, charity as their religion dictates it is achieved as a single church body, and aesthetic experience when as individuals experiencing art together they share a distinctively different communal experience of beauty then when going to the museum all alone.

Now imagine Jack and Alex decide their affection for one another should be more fully expressed by raising a child together, but as a homosexual couple they cannot realize that human good *tout court* and they also do not wish to adopt a child from another with whom they share no strong friendship based on mutual interests, loyalty and even love. Suppose also that Jack and Alex discuss their deeply held desire to share, deepen and advance their love with one another and with a new human life by having a biological child that is (one of) their own, but they also do not wish to miss out on the interpersonal reality of conceiving a child by instead merely legally contracting with an unknown or anonymous surrogate mother with whom they would share but an episodic and lonely experience as laboratory bio-parents. Perhaps they even add that they find it very impersonal and too "reductive" of their possibilities of well-being the way some surrogate mothers use private agency databases to promote their attractive features—i.e., young, pretty, Ivy League degree—for they fear having a mere bio-mother who provides options from which to choose, like a restaurant menu, would alienate them from the very child they wish to conceive in love and devotion. Instead they wish to preserve a strong interpersonal relationship with the mother of their child and they ask Sally to conceive and bear their child using Alex's sperm—perhaps the two even engage in intercourse to make this possible—and Sally then plays a large part in the child's life as a main but secondary caregiver. Now suppose Sally's bio-child, named Isabelle, lives primarily with Jack and Alex who raise her on their own with Sally playing a significant interpersonal part of the family. However, as Sally does not live with Isabelle she is not as involved as many other parents living in the house would otherwise be.

At least as so constructed—and this is not just a fictional scenario in today's world[12]—it is not clear why Jack and Alex could not also satisfy many, perhaps all, of Tollefsen's criteria for creating and rearing a child within a family that maintains the other-related, interpersonal realities of "marriage" and friendship. Jack and Alex (and even Sally) seem to have interpersonal communion that extends beyond their own narrow range of selfish desires, if perhaps only because they provide, care for and love Isabelle day-in and day-out. Now if Jack and Alex decided that their love for their family would be more complete if Isabelle had a brother or sister but Sally did not want again to be a mother, suppose Alex was asexually cloned and the new child in the family were named Alexandra. It is true by definition that Jack and Alex do not share the goods inherent to a heterosexually grounded marriage marked by the heterosexually interpersonal reality of sexual congress that lovingly produces a child.

But even if Alex autonomously procreated Alexandra his own well-being has many (other and additional) expansive possibilities—from one who is Sally's friend and appreciates art with Jack and engages in community projects with them both and is as well a parent to Isabelle and Alexandra—that do not remove him from Others with whom he has interpersonal communion. In fact, while Alex's creation of Alexandra (with some help from the lab) may look like "masturbating for children" (Tollefsen, p. 23), if Jack and Alex had their children because they wanted to share their love in their "marital" relationship—much like (but of course not exactly like in every way) heterosexual couples might claim such affective and embodied reasons for having children—and if they raise Isabelle and Alexandra in a loving and caring home, it is unclear (without begging the question) why their interpersonal communion day-in, day-out, as parents and friends would alienate them from both Isabelle and Alexandra.[13] So, even if the dysfunctional lives painted in Aldous Huxley's vision of a bleak future in *Brave New World* certainly suggests (nearly) exclusive asexual reproduction by scientific means sacrifices the natural interpersonal goods of friendship and family (Tollefsen, pp. 23–24), it is unclear why short of Huxley's tragic and dystopic future that Alex and Jack's actual non-marital and asexual reproduction—in part or whole—would necessarily sacrifice all or many of the substantive goods inherent to families who love and respect their children and raise them to care for one another.

And this would also seem to hold true for the form of religion that Alex and Jack, and even Sally, Isabelle and Alexandra might practice together in order to gain Tollefsen's basic good of "transcendent meaning." Alex, Jack and Sally might claim that their basic goods are part of a complex human nature they practice in *some* socially unorthodox ways. However, as they join their Christian church brothers and sisters on the weekend to build a Habitat for Humanity house they might claim they are harmonizing their wills with others and with their Christian source of "transcendent meaning" by working together and giving charitably of their time and energy for the less fortunate children of God. Alex, Jack and Sally could even add that by showing Isabelle and Alexandra how Christians treat the least of their brothers they are following in Christ's footsteps. All three of these adults might even insist that their "practical nature" specifies and limits the "basic goods that constitute the horizon" of their "possible flourishing" to a "finite, relatively small, and permanent" set of goods made possible by their autonomous "deliberation, choice and commitment." That is, Alex, John and Sally might acknowledge that they are so constituted as humans in social situations with limited time and resources so that they can only realize some but not all of the goods inherent to friendship and health and art that serve individual as well as social purposes. However, this alone and without more does not show why their same-sex family arrangement fails to realize the broader range of fully human goods that a faithfully married heterosexual couple realize when attending a museum opening, building a Habitat house or attending religious services on Sunday—and when raising children in a loving home.

Of course it might be that Alex, Jack and Sally living in New Jersey could not in good faith engage in Roman Catholic religious worship to achieve "transcendent meaning" if they did not honestly foreswear and seek forgiveness for their extra-marital sexual practices and lifestyle. But even if this is true as a matter of Roman Catholic theology it is not necessarily true of all kinds of Christian religious practice, to say nothing of other forms of religious worship. So if it is true that Alex, Jack and Sally could achieve a broad and full range of basic goods like art and friendship and knowledge while living in a same-sex family arrangement, it is not clear if or why the "transcendent meaning" of religion *as such* is for them thus foreclosed. So as for Lee then also for Tollefsen it might be the case that a more robust account of Catholic theology and religious practice would make clear why an unorthodox family arrangement with same-sex "marital" partners prevents them from securing the "transcendent meaning" of religion and perhaps also from achieving the basic goods of true friendship *and* rich aesthetic experience *and* intrinsically rewarding knowledge—for individuals alone and people communally in groups—but that would take a longer and more detailed argument than presented here.

But despite this potential pitfall, Tollefsen offers four additional points that are both imaginative and interesting for teachers and students alike for extending natural law theory into the modern realm of technological change in ways that Kantian, Utilitarian and virtue ethics just might not be able to manage.

The second point worth noting is Tollefsen's claim that beings of greater intelligence achieved by combining biological and technological tools could violate the "separateness of persons" by combining silicon selves with biological ones, but it is that very separateness itself which limits friendship and so enables its basic goods (Tollefsen, p. 24). So even if friendship is limited by being apart from the Other it is thereby enabled, too, since that separateness empowers us to reach out to the Other to build a mutual association based on similar interests and values as well as by promoting the Other's good.

Third, Tollefsen claims modern-day cognitive "enhancers" such as Prozac and Ritalin when taken not by the sick but by the healthy threaten our ability to make our own lives and choices by serving as a "substitute for the labor" of working harder and better at self-constitution, or at integrating our emotional and social lives on our own (Tollefsen, pp. 24–25). Likewise using technology to "download a character change" and "new identity" might make self or society happier but they would also deprive us of achieving the basic goods as constituted by acts of will (Tollefsen, pp. 26–27). As an example, again consider friendship. A genuine and authentic friendship "cannot be downloaded. . .or installed" because we are then deprived of its basic good as constituted by "human choices, and commitments" (Tollefsen, p. 26).

Fourth, some advancements in neuroscience might "make us more" "susceptible to physical pleasure" than we are today and might thereby make "temperate choices more difficult," thus threatening the goods of human autonomous choice we could achieve (Tollefsen, p. 28).[14]

Fifth, Tollefsen begins by repeating the claim that such so-called technological and medical enhancements to the body or mind, if they do not impair our "relationship to certain goods" or to act autonomously, do not "constitute *any* genuine change in human nature" and thus are apparently morally permissible as well (Tollefsen, p. 28; emphasis in original). However, if some research in cloning, genetics and robotics requires experimentation on human embryos and thereby violates rights that "accrue to individuals in virtue of what they are"—e.g., a right to life that "belongs to us as long as we exist"—such research itself (but not necessarily the enhancements) would thereby be immoral for violating those basic rights.

Just as they might benefit from studying the natural law account of friendship by Lee, teachers and students of ethics alike might profit from examining Tollefsen's comparably robust natural law assessment of modern-day medical and technological research. At first glance such teachers and students might think that, say, a Kantian or Utilitarian account would be easy to offer and quite useful, too, for analyzing such present-day human enhancements and determining their moral status. After all, it would seem that any rational agent could consent to their use to enhance our intelligence and certainly the individual and social benefits of such enhancements might, on balance, outnumber and outweigh their costs. However, as in assessing the ethical status of friendship above in Part one these two traditional moral theories can falter in these cases of medical enhancements as well here in Part two.

For starters a Utilitarian might well defend cloning or any cognitive "enhancers" if the subsequent experiences of sentient agents produced a net gain in individual and social pleasure over pain—and even if the basic human goods constituted by free autonomous choices in marriages and friendships were as severely damaged as Tollefsen maintains. Moreover, it is certainly more than plausible that a Utilitarian of any stripe might defend all such research in the face of the destruction of human embryos just so long as an individual and social net quantitative and qualitative calculus of pleasures came out just right.

But it is quite hard to know when that is just right. Since the difficulty Utilitarians have more generally in making clear and confident empirical predictions are sufficiently robust when merely deciding whether to lie to Aunt Mame about her Christmas fruitcake, one can only imagine how difficult it will be to balance the net gains against the costs of, say, asexual cloning and cognitive enhancers and character identity program downloads when in the mix one must off-set against the benefits of increased intelligence and reduced personal anxiety and new individual personalities the very real risks of loss of (or utter transformations to) identity, friendship and marriage. Even a sophisticated rule-utilitarian might well be at a total loss for which rules-of-thumb might provide even rough guides for helping to adequately determine the balance of pleasures and pains, to say nothing about just how Millian qualitative utilitarians might plausibly defend the experiential reasons that the so-called competent judges would not trade the intrinsic values inherent to friendship and marriage for any amount of physical or mental pleasure produced by cognitive enhancers and entire character downloads. After all, it might well be that nearly all people who have had physical and mental pleasures would jump at the chance to download a new personality and get a cognitive enhancement upgrade to avoid the anxiety of modern life.

For its part Kantian ethical theory might avoid these problems while also giving voice to our intuition that merely using human embryos in medical tests to produce such alleged benefits is morally wrong. A Kantian argument could be offered to construct an objection to any research that merely instrumentalized such (potential or actual) bearers of rational life such as human embryos. For example, a Kantian might suggest no rational agent could now consistently consent to have been merely used (and thus killed) in an experiment to make possible future genetic or robotic enhancements from which he could not, by definition, benefit and which earlier in time would have made life now (for him) impossible.[15] Likewise, for a Kantian downloading a character or identity or not choosing to constitute our very selves by working harder (Tollefsen, p. 26) might bring about valuable changes but might also prevent us from exercising the acts of will necessary for expressing the rationality characteristic of acting from duty alone and not from inclination.

On the other hand, it is not clear that Kantian ethical theory would be really four-square opposed to all human use of cognitive "enhancers"—whether drugs like Ritalin taken simply for improved cognitive performance or computer-chip implants intended to improve intelligence—since it is not so clear that such use fails a test of universalizability or treats either oneself or rational second-parties as mere means. Contrary to Tollefsen's empirical claim (p. 29), it might be possible to produce such enhancers and chips without any human embryo research and a Kantian could argue that a free choice to take such enhancers would only further empower our rational agency and so not deepen our reliance on inclination and passion. Thus, it is not clear a Kantian can or should object to enhancers or chips or identity downloads as violating universality and as violating proscriptions against merely using rational agents, but might even instead defend them as being consistent ways of exercising our agency and respecting another's autonomy.

Despite the potential counter-intuitive moral endorsement of consensual non-marital sexual acts between people of the same or different sex—about which both authors would surely have more to say—it would appear that Tollefsen's natural law account, like Lee's, might be welcomed by ethics teachers and students alike as more solidly grounding moral assessments of a broad range of human practices and activities and behavior, from friendship to cloning and performance-enhancing drugs, than other traditional moral theories can manage. If that were possible, between the Scylla of ethical relativism and the Charybdis of postmodern culture some reasonable and coherent ethical consensus on complex moral issues might well be possible after all by turning to such rich and robust modern-day accounts of natural law. Perhaps all that is wanting now is for ethics teachers—or maybe book publishers!—to devote as much time in their classes and texts to accounts of natural law from philosophers like Tollefsen and Lee as they do to other traditional ethical theories whose challenges are as often overlooked as the normative power of natural law theory demonstrated in these articles is ignored.

Notes

1. The ones here are drawn from Chapters 2 and 3 of James and Stuart (2007a). I will have more to say about that text below.
2. Rorty's suggestion that we avoid the conceptual bankruptcy of defending philosophical foundations of ethics and instead more pragmatically change the topic—not answering the charge of relativism but evading it by redefining the moral issues in linguistic and historical ways—might be something a much more sophisticated student could offer (1990, pp. 54–55). And perhaps even here there is little that is clearly false but instead misplaced about trusting metaphoric and cultural considerations to secure the moral case that governments *should* reduce suffering while letting individuals pursue

their private lives—which for Rorty "seems. . .pretty much the last word" (p. 63). That is, it may be difficult to show what is false but instead impractical about Rorty's postmodern turn away from philosophical truth to communicative rhetoric. That said, once the argument turns this way any use of words like "true" or "better" or "good" might, at best, equivocate on clear meanings for speakers and listeners as Rorty engages in creative redefinitions of ethical terms, thus making his very project hard to follow in the first place. (My thanks to Mark Cherry for suggesting the multiple accounts of relativism at play in these discussions.)

3. I owe the suggestion about "mourning" these old selves to philosopher Martin Benjamin.
4. In fact, the numbers have nearly reversed from 1968 to 1996. In the late 1960s more than 80% of new freshman indicated that "developing a meaningful philosophy of life" was "essential" or "very important" while "being very well off financially" ranked fifth or sixth with less than 45% endorsing that as a very important or essential goal. By 1996 the financial goal had moved to the top with 74.1% endorsing it, while a meaningful philosophy of life dropped to sixth place with 42.1% endorsing that goal (Astin, 1998, p. 124). While this study suggests these numbers are correlated with television watched by freshman (p. 125), this reversal may reflect other trends and causes. For example, colleges might be now opening their doors to less affluent students, so these statistics may track beliefs determined by family income more than independently constructed ethical convictions. Likewise such a reversal may more directly be determined by political considerations such as the Vietnam War influencing students' draft-deferred commitments to a college focusing on a "meaningful philosophy of life" in the 1960s when contrasted with other commitments to college in the 1990s. This appears more likely if more students in the 1990s are bearing the costs of college education and thus more likely to frame college goals in terms of financial well-being associated with paying off grants and loans needed to go to school in the first place. (My thanks go to Mark Cherry for suggesting similar explanations.) However, it seems also plausible that this reversal reflects a postmodern culture's inability to make philosophy in general and ethics in particular an important life's pursuit, especially in the face of other economic and political influences that might even benefit from a more thorough philosophical and ethical analysis to determine what make such influences valuable in the first place.
5. See "College Kids Choose Adderall Over Ritalin For Illicit Use," (2007), citing joint study of researchers at Northeastern University and the University of Michigan.
6. Foot actually claims that the virtues of charity and justice do not "correspond" to "a desire or tendency" to reign in but rather to a "deficiency of motivation" (1998, p. 170).
7. The comedian Bill Maher and the writer Susan Sontag stirred controversy by suggesting a few years ago that the 9/11 terrorists were courageous for facing their fear as they engaged in their horrific homicidal acts. On Foot's account they would both be wrong.
8. My thanks to Mark Cherry for suggesting how the virtue of truthfulness, even and especially when embedded in different traditions, requires more complex decision-making procedures than one might otherwise think.
9. The Rachels companion volume to the *Elements* book mentioned above collects some primary texts, including only two entries on natural law in its 31 chapter selection. One includes four and half pages of ill-chosen Aquinas (Chapter 5) and the other a longer reading from Burton M. Leiser (Chapter 13) on homosexuality. However, the Leiser entry mostly covers the very view that Lee rejects here.
10. So much for the nearly five pages on natural law in the Rachels *Elements* volume that claims for this view that "[m]oral rules are now viewed as deriving from the laws of nature" (p. 60).
11. Lee's footnote about "self-integration" cites Germain Grisez, though these passages cited here come from Robert P. George's "A Clash of Orthodoxies" (1999).
12. For comparable accounts see "Your Gamete, Myself" (Orenstein, 2007) and "Gay Donor or Gay Dad?" (Bowe, 2006).
13. There would no doubt be much social disapproval expressed about this unorthodox family—even in New Jersey—that might lead John and Alex to feel alienated from Isabelle and Alexandra in ways few heterosexual parents would ever experience. But even if this were true such social causes of alienation are not quite the same as those hard-wired into our practical natures that Tollefsen is defending here.

14. Tollefsen earlier made a related point about the moral cost of asexual cloning if that reproductive technology changed our human animal identity to such a degree that, as individuals or as a species, we could no longer reproduce sexually. In this case Tollefsen argues our "relationship" to the basic goods of marriage and reproduction would be cut off (Tollefsen, p. 23).
15. Harry J. Gensler argues something like this in "A Kantian Argument Against Abortion" (1986). However, the force of this argument rests on having certain consistent sets of psychological desires based on one's assent to certain hypothetical claims about when and where one would have conditionally accepted injury and death at earlier stages in life. This may be a perfectly sensible Kantian approach to abortion. However, its complex formula of counterfactual logical assertions resting on certain psychological states not only seems to beg a number of questions but it leaves one cold: This is how an agent is supposed to come to the conclusion that abortion is morally wrong? That is, by engaging in counterfactual logical assertions premised on certain psychological states that I do or should have? Comparing this to Tollefsen's account above provides a stark contrast

Bibliography

Astin, Alexander W. (1998). The changing American college student: Thirty year trends, 1966–1996. *The Review of Higher* Education, 21(1).
Bowe, John (2006). Gay donor or gay dad? *New York Times*. 19 November.
Foot, Phillipa (1998). Virtues and vices, in R. Crisp and M. Slote (eds.), *Virtue ethics*. New York: Oxford University Press, 163–177.
Gensler, Harry J. (1986). A Kantian argument against abortion. *Philosophical Studies*, 49, 83–98.
George, Robert P. (1999). A clash of orthodoxies. *First Things*. August/September, 33–40. 2006–2007. [On-line]. Available: www.firstthings.com/ ftissues/ft9908/ articles/ george.html (28 September 2007).
James, Rachels and Stuart, Rachels (2007a). *The elements of moral philosophy*, 5th edition. Boston: McGraw Hill.
James, Rachels and Stuart, Rachels (2007b). *The right thing to do: Basic readings in moral philosophy*, 4th edition. Boston: McGraw Hill.
MacIntyre, Alasdair (1984). *After virtue: A study in moral theory*, 2nd edition. Notre Dame, Indiana: University of Notre Dame Press.
Medical News Today. (2007). College kids choose Adderall over Ritalin for illicit use, 21 October. [On-line]. Available: www.medicalnewstoday.com /medicalnews.php?newsid =55427 (22 August 2007).
Orenstein, Peggy (2007). Your gamete, myself. *New York Times*. 15 July.
Rorty, Richard (1990). *Contingency, irony, solidarity*. Cambridge: Cambridge University Press.

Part II
Human Goods and Human Flourishing: Revitalizing a Fallen Moral Culture

Quid Ipse Sis Nosse Desisti

Douglas V. Henry

Few instances of equanimity in the face of personal ruin inspire wonder as the final days of Anicius Manlius Severinus Boethius. *The Consolation of Philosophy* movingly narrates Boethius's reversal of fortune, relating his fall from high political favor into the disgrace of exile and execution as a traitor.[1] A genuinely wise man—inspired to public service by a "desire for good" and by reading Plato (1.pr.4)—he feels the full force of destitution and wavers momentarily before coming to himself. Through the persistent and patient nursing of Lady Philosophy, Boethius is reminded that "no sudden change of circumstances ever occurs without some upheaval in the mind; and that is why you, too, have deserted for a while your usual calm" (2.pr.1). Unwilling to see one of her most discerning disciples despair, Lady Philosophy undertakes first to diagnose and then to treat Boethius. As book one concludes, she completes her diagnosis. Her desolate disciple knows that the world is not the result of "haphazard and chance events" because it is governed by God, but he does not know how God rules creation. He remembers that God is the "source from which all things come," but because his "memory has been blunted by grief" he no longer recalls "what is the end and purpose of things." And tellingly, he acknowledges that "man is a rational and mortal animal," but insists assuredly that humankind is "not something more." In response, Lady Philosophy says, "Now I know ... the major cause of your illness: *quid ipse sis nosse desisti*" (1.pr.7.17). *You have forgotten what you are.*[2] "And so," Lady Philosophy asserts, "I have found out in full the reason for your sickness and the way to approach the task of restoring you to health."

A more fitting judgment of us and of our time, at the end the modernity, could scarcely be found. We too share in Boethius's malady, having forgotten what we are. Perhaps unlike Boethius, our failure of self-understanding is no mere blunting of memory through grief-worn care, but instead constitutes a willful giving over of, a desisting from, knowledge of what we are—and more. While we share with Boethius the failure to know what we are, our failure is deeper than his, for

D.V. Henry (✉)
Great Texts Program, Baylor University, Waco, TX, USA
e-mail: Douglas_Henry@baylor.edu

M.J. Cherry (ed.), *The Normativity of the Natural,* Philosophical Studies in Contemporary Culture 16, DOI 10.1007/978-90-481-2301-8_6,
© Springer Science+Business Media B.V. 2009

only seldom do we moderns profess the modest knowledge that he embraces when acknowledging God as Creator, the world as meaningfully ordered, and humankind as rational and mortal. If Boethius's deficiencies are "grave" and "lead not only to illness but even death," what peril indeed must we, at the end of modernity, confront? Can philosophy offer us consolation or cure?

I want to argue—in light of the insights that Boethius offers in his wildly popular medieval text—that philosophy can both console and cure us.[3] When it succeeds, it is not through the false therapies that, as Lady Philosophy warns Boethius, "slay the rich and fruitful harvest of Reason with the barren thorns of Passion" (1.pr.1). Instead, philosophy succeeds where therapeutic drivel cannot precisely when it discerns what we are within the context of our source and our end. By so doing, knowledge of what we are offers morally significant direction for our lives. And lest we dismiss Boethius's confidence as the overblown buoyancy of a broken down old man who imagines a shapely goddess of truth, let me insist that philosophy need not be any less efficacious when we encounter it in its more pedestrian forms.

I will offer my favorable estimate of Boethius's movement from dispossession to consolation by way of lively engagement with a couple of leading sources of philosophical discontent with the idea of human nature. These two means of erring in one's philosophical anthropology come by way of scientific rationalism and misologistic nihilism. Ancient forms of both scientific rationalism and misologistic nihilism variously enervate Boethius, giving rise to his uncertainty about what he is and his moral torpor; likewise, contemporarily clad manifestations of scientific rationalism and misologistic nihilism beset us in late modernity, giving rise to our uncertainty about what we are along with our own troubling species of moral ambiguity.[4] It thus should be clear that I occupy a definite position in relation to the central issues of this book, namely that human nature rightly understood is morally normative; that modern philosophy is discontent with the idea of human nature and is, *ipso facto*, lacking sound knowledge not only of what we are but of how we should live; and that philosophy ideally in theory and practice offers us—like the beautiful specter of wisdom that Boethius beholds—the prospect of true consolation and cure.[5]

1 Scientific Rationalism

In book one of the *Consolation*, Boethius, forlorn and caught in the grip of despair, offers a succinct account of what we are. The dialogue between Lady Philosophy and Boethius unfolds straightforwardly:

> ...I want you to answer this too: do you remember that you are a man?
> Why shouldn't I? I said.
> Can you, then, tell me what man is?
> Are you asking me if I know whether man is a rational and mortal animal? I do know it and I acknowledge that that is what I am.
> Are you sure you are not something more?
> Quite sure.

So it is that Lady Philosophy diagnoses "the major cause" of Boethius's malady. He understands human beings to be nothing more than "rational and mortal animal[s]" and assuredly denies the possibility that we are "something more." Thus does Boethius voice an ancient, deflationary account of human nature and presciently anticipates the staying power of reductive, "nothing but" philosophical anthropologies, sustained in late modernity in the form of scientific rationalism.

By "scientific rationalism" I mean the tendency to privilege forms of reasoning regarded as scientific to the exclusion of other valid forms of reasoning, be they philosophical, theological, ethical, or what-not. Persons holding to this view believe that all phenomena best can be accounted for in scientific terms. Although allowances sometimes might be made for psychological, literary, or even religious accounts, they inevitably are relegated to a second-class status, to be refined by, reduced to, or completed by science. Scientific rationalists insist upon the epistemological superiority of their general way of understanding the world to all other ways. Crucially, this position is motivated in most cases by a certain kind of metaphysics: scientific naturalism. Scientific rationalism is most comfortably embraced in the company of a scientific naturalism that holds that, in the end, only the natural objects studied by science exist, which is to say that only physical properties and things exist. As John Post puts it:

> According to a number of influential philosophers, the sciences cumulatively tell us, in effect, that everything can be accounted for in purely natural terms. . . . The worldview this entails, according to many, is *naturalism*: Everything is a collection of entities of the sort the sciences are about, and all truth is determined ultimately by the truths about these basic scientific entities (1991, p. 11).[6]

Scientific rationalism, underwritten by a naturalistic metaphysics, tends to give expression to an outlook that is totalizing and reductive, despite efforts to soften these sharp edges.[7] It is totalizing in that it purports to explain everything, and it is reductive in that any alternate explanation of things ultimately can be "reduced" into its own terms.

For a scientific rationalist, knowledge of what we are, of human nature, inevitably means knowledge gotten by science, using its methods, employing its terms, and staying within its purview. Devotees to this outlook suppose that science can describe human nature, without remainder, in the terms set forth by its sub-disciplines. A contemporary scientific rationalist's account of what we are might therefore begin broadly with sociology, but from that rough-grained picture of human nature would work down through psychology, biology, and molecular biology to chemistry, and from there would drill right on down to the way things "really are" as described by physics. Robust, morally normative accounts of human nature cannot but suffer loss on such terms, for if the "thorough-going reductionist" is right in claiming that "[u]ltimately all is physics," then as John Polkinghorne puts it, "[e]verything else is nothing but an epiphenomenal ripple on the surface of a physical substrate . . ." (1986, p. 86). Even when "physics envy" and the extreme reductivism it prompts do not win the day, knowledge of what we are nonetheless suffers diminishment in the hands of scientific rationalists.

1.1 Scientific Rationalism: Human Nature in Dennettian and Boethian Perspective

Before returning to Boethius's *Consolation*, let me identify a fine though flawed effort made by a leading scientific rationalist to develop a morally normative understanding of human nature. In *Darwin's Dangerous Idea*, Daniel Dennett, among other things, evaluates several of the morally normative accounts of human nature typical of scientific rationalism. Despite sharing with them an extraordinarily high view of science, he frankly sees problems with the conclusions reached by kindred spirits such as B. F. Skinner, E. O. Wilson, Michael Ruse, and Richard Alexander. As he puts it:

> Another part of our inquiry into human nature, as a naturalistic basis for sound ethical thinking, would begin with the undisputable fact that we human beings are products of evolution, and consider what limitations we are born with and what variations there are among us that might have ethical relevance. Many people apparently think that ethics is in deep trouble if it turns out that human beings aren't, as the Bible tells us, just a little below the angels. . . . But there are more reasonable visions that are also jeopardized by the discoveries of the scientists (not just evolutionists) (1995, pp. 481–482).

Dennett's project, in short, is predicated upon the unrealized aspiration to escape the jeopardy into which "the scientists" have placed ethics, including the efforts of the Skinners, Wilsons, Ruses, and Alexanders of the world. Finding their views wanting but their aims admirable, Dennett offers an alternative account of human nature and a concomitant ethical outlook. His account is self-admittedly reductive, like theirs, but unlike theirs it is not "greedily" reductive.[8] He begins by arguing that the so-called naturalistic fallacy (whereby it is impermissible to derive an "ought" from an "is") cannot mean that ethics is altogether disconnected from the facts about reality in general and human nature in particular. He asks:

> If "ought" cannot be derived from "is," just what *can* "ought" be derived from? Is ethics an *entirely* "autonomous" field of inquiry? Does it float, untethered to facts from any other discipline or tradition? Do our moral intuitions arise from some inexplicable ethics module implanted in our brains . . .? (1995, p. 467).

Dennett maintains that "ethics must be *somehow* based on an appreciation of human nature—on a sense of what a human being is or might be, and on what a human being might want to have or want to be. . . . No one could seriously deny that ethics is responsive to such facts about human nature" (p. 468).[9] He wants to take ethics seriously as an important feature of human experience and he wants to make sense of ethics on the basis of what one can discern about human nature—provided that one's understanding of human nature is rigorously scientific, subject to the methods, terms, and limits of science. For this reason, he perhaps ranks among the best that scientific rationalism can offer in terms of a morally normative account of human nature.[10]

That said, Dennett's account falters. It falters just in the way that Boethius's account falters. But before explaining why both versions falter, let me address what each of them gets right. Recall that Boethius understands human beings to be "rational and mortal animal[s]." Surely he is correct. We are like other mortal animals in our possession of nutritive and sensitive faculties, and we share in common with all animals a natural cycle of birth, life, and eventual death. We come to be and we pass away, and in between our beginning and our ending we experience and interact with the world around us. Shifting to the contemporary terms used by Dennett, "we—*Homo sapiens*—are one of the species over which evolutionary theory reigns" (1995, p. 335), and we are born, we live, and we die subject to the same biological laws that unite us with other forms of animal life. Yet, Boethius recognizes, we are unlike other animals in our capacity for reason. Even if we generously suppose that human reason differentiates us from other animals only in degree rather than kind, no one convincingly maintains that there are not real differences. Dennett is perfectly willing to accept the point. He writes, "We are not like other animals; our minds set us off from them" (p. 370).[11] My point is that the belief that humans are rational and mortal animals is right insofar as it goes, and good common sense deserves to be acknowledged. The problem is that this account does not go far enough; it does not embrace the "something more" which Lady Philosophy regards as indispensable for a complete account of human morality.

In Boethius's version, the morally ordered cosmos he discerns does not ultimately hold together. So long as Fortune favors him, he experiences a happy congruity among the goods of the intellect, the justice and nobility of his ways, and the extrinsic rewards of his virtue. And so long as everything works out so obviously well, perhaps knowing that one is a rational and mortal animal is enough; it is at least enough to sustain the illusion that reason resolves life's puzzles without remainder into morally meaningful and satisfactory solutions. But Boethius learns that Fortune does not play favorites, at least not forever: "While with success false Fortune favoured me / One hour of sadness could not have thrown me down, / But now her trustless countenance has clouded, / Small welcome to the days that lengthen life" (1.m.1). His bitter, hard-won experience teaches him that no one is guaranteed the happy congruity of rationally underwritten virtue and blessedness in this mortal life. Worse than that, he comes to recognize that plenty of perfectly decent people are afflicted with manifest injustice and untold misery, and that plenty of perfectly horrible people enjoy lives of comfort and apparently undisturbed happiness—a circumstance he decries as "monstrous" (1.pr.4) and laments as "beyond perplexity and complaint" (4.pr.1).[12]

Put another way, the persuasive deliverances of *sola ratio* on behalf of a morally well ordered life, at least for Boethius, fall on hard times in the face of his existentially significant encounter with a rationally impenetrable instance of the problem of evil. And so, he grieves in the marvelous fifth poem of book one:

> . . . All things obey their ancient law
> And all perform their proper tasks;
> All things thou holdest in strict bounds, -
> To human acts alone denied
> Thy fit control as Lord of all.
> Why else does slippery Fortune change
> So much, and punishment more fit

For crime oppress the innocent?
. . . Bright virtue lies in dark eclipse
By clouds obscured, and unjust men
Heap condemnation on the just;
No punishment for perjury
Or lies adorned with speciousness.

In a world in which reason regularly beholds "bright virtue. . .in dark eclipse," why should one be moral? If the sensibly and intelligibly apprehended cosmos does not acknowledge, indeed underwrite, the merits of rational and mortal human nature, then how one should live goes up for grabs, becoming radically underdetermined by the way things are. Boethius sees the gap, and he petitions God to amend a rational and moral affront that otherwise appears without remedy. In the end, Boethius embraces Lady Philosophy's wisdom by accepting that human beings are "something more" than merely rational and mortal animals. Specifically, the something more that he accepts involves discerning that human nature, rational and mortal as it is, properly finds its meaning within the framing context of its origin and end in God. Apart from God's work in providentially guiding the affairs of the world, Boethius sees that reason suffers the dire threat of bankruptcy in the face of evil. Apart from divine superintendence of human life, Boethius agonizes over the ways in which evil devalues the virtuously disposed deliverances of reason. He understands that in the absence of divine providence directing, ordering, and ruling the world, human nature's rationally underwritten and morally normative propensities cannot assuredly provide morally significant direction for life. For these reasons, he accepts that what defines human beings are not merely the qualities of animality, mortality, and rationality, but also the quality of having been created providentially by and for God.[13]

In similar fashion, Dennett's evolutionarily grounded account of the normative dimensions of human nature stumbles, though with a significant difference: he does not concede the inadequacy of his account as Boethius does. Dennett's evolutionary account gets much right. We are, he believes, "finite, time-pressured, heuristic searchers for ethical truths" (1995, pp. 493, 505). The evolutionary unfolding of a world in which human beings, marvelous that we are, exist as moral agents, and in which we are endowed with the capacity to think our way through moral quandaries, may be reasonably accepted. Fully valuing the finer points made possible through Dennett's work, the problem nonetheless remains: the world is not consistently friendly to rational virtue, or, put in the parlance toward which he gravitates, being a good person does not always have "survival value."

In the face of this reality, Dennett is inclined in two directions. The first is to acknowledge that:

> every day, while trying desperately to mind our own business, we hear a thousand cries for help, complete with volumes of information on how we might oblige. How on Earth could anyone prioritize that cacophony? . . . By and large, we must solve this decision problem by permitting an utterly 'indefensible' set of defaults to shield our attention from all but our current projects (pp. 509–510).

That is, Dennett appears to allow the worst kind of derivation of an "ought" from an "is," one that succumbs to the kind of "greedy reductionism" that he is intent on avoiding. For on his account, Nature privileges our ordinary moral *modus operandi*, and it "justifies" (indefensibly!) our *status quo* rules for resolving moral questions. Talk of rights, for example, is "nonsense upon stilts, but good nonsense—and good only because it is on stilts, only because it happens to have the 'political' power to keep rising above the meta-reflections" (p. 507). Likewise, he writes, "'rule worship' of a certain kind is a good thing .... It is good not because there is a certain rule ... which is provably the best, or which always yields the right answer, but because having rules works—somewhat—and not having rules doesn't work at all" (p. 507). Dennett sees no rational justification for moral rules, but asserts their value. Dennett regards rights-talk as nonsense, but to be welcomed because of its "political" power as a "conversation-stopper" that encourages self-critical reflection. And behind it all, supposedly, Mother Nature exhibits "genius" in its "blind, purposeless ... recognition of a good (a better) thing" (p. 511). It is difficult to understand, much less to celebrate, anything like *blind, purposeless genius*—an oxymoronic expression if ever there was one—and not a few reviewers confess an inveterate impulse to dismiss such florid declamation as the bombast that it is.[14]

The second direction intimated by Dennett is that of holding out for more time, better evidence, or superior rational insight. In a remarkable circumlocution, Dennett writes: "no mere brute fact about the way we are built is—or should be—entirely beyond reach of being undone by further reflection" (p. 506). He seems to believe that given a sufficiently long period, scientists will be able to solve the riddle of human existence, and with it make clear sense of what we are, together with the morally normative implications of what we are. Yet the temptation, I would argue, is to overestimate the possibilities of scientific theorizing and the rational control over life of which we are thereby capable. The temptation is to suppose that we can domesticate Fortune and have her do our bidding, thereby putting the cosmos in our grip and not the other way around. Where Dennett does not baptize the irrational, unjustifiable, but evolutionarily derived status quo as morally apropos, then, he instead holds out for an ideal account at a future point that offers meaning where none is now apparent.[15]

In short, neither of Dennett's responses is persuasive. The first response promises to abandon reason without really admitting it; the second response just promises. Given the obvious merits of rationally ordered and morally integrated lives, nonsense upon stilts and "indefensible" moral defaults will not do. And given the availability of an alternative account of the morally normative character of human nature, the one that Boethius's Lady Philosophy is prepared to offer, for instance, I do not see the merit of waiting indefinitely for an ideal account always receding into an infinite future, and based upon a promissory note that science will figure it all out. Fraud and hot check writing are (or used to be) punishable crimes; if public intellectual life were ordered analogously to public economic life, Dennett would be guilty of versions of both offenses.

To be sure, not all modern scientists are of the class to which Dennett belongs, and I must emphasize that some modern students of science fully appreciate the need for "something more." So it is that the medical doctor and critically acclaimed author Walker Percy, about whom I will say more below, grappled intensely with the same dicta derived from modern science and realized, in the spirit of Boethius and his Lady Philosophy, that the way of the scientific rationalist leaves much wanting. "After you learn everything you can at Columbia about what it is to be a human being," he reflected, "there is something awfully important left over" (Percy, quoted in Wood, 1988, p. 146). In his fine treatment of Percy, Ralph Wood explains that:

> about our unique condition—what we are as singular individuals—[science] cannot utter a solitary word. This is the drastic limit, Percy discovered, set upon all scientific knowledge. Science can comprehend the meaning of life within its own horizon, but it is powerless to deal with the transcendent Reality—if there be any such—that impinges upon us. Percy's melancholy recognition of this fact left him feeling strangely empty (1988, pp. 145–146).

Percy's melancholy recognition of science's inability to address human meaningfulness occurred early in his life, amidst and after his study of medicine at Columbia University. As he later on expressed it, he came to terms with the import of Kierkegaard's famous dig at Hegel—except in application to latter-day scientific rationalism rather than to Hegelianism. "After twelve years of a scientific education, I felt somewhat like the Danish philosopher Søren Kierkegaard when he finished reading Hegel. Hegel, said Kierkegaard, explained everything under the sun, except one small detail: what it means to be a man living in the world who must die" (Percy, 1991, p. 188). What that means, what that involves, Percy saw, entails

> recognition of a standard of human existence wholly different from that by which we judge the flora of Australia or the ape population of the Congo. It means that there is being proposed as the central criterion of man's well-being the very thing most detested by the biological method: a value scale of rightness, authenticity; in short, a concept of human nature and what is proper to it (1991, pp. 256–257).[16]

Percy was an ardent fan of the manifold accomplishments of modern science, and he had no desire for a general return to Aristotelian teleology, with all of its inadequacies as a ground for natural science. What he insisted upon, nonetheless, was the need for science to recognize its proper limits and not to claim the capacity to explain more than its assumptions and methods made possible. Above and beyond science rise the great and recurring questions of human significance, and the capacity to recognize these questions and to seek answers to them is partly constitutive of "human nature and what is proper to it" (though not exhaustive of it).

In response to all reductive, nothing-but accounts of human nature—be they found in the form of Boethius's unwitting forgetfulness, Dennett's self-assured triumphalism, or the young Percy's melancholy recognition—Lady Philosophy asks, "Are you sure you are not something more?" Are we merely rational and mortal animals and nothing else? To the extent that scientific rationalism withholds answer, or confidently answers in the negative, or morosely broods upon the strange emptiness of human existence, it shows forth one way in which of we moderns it may be said, *quid ipse sis nosse desisti.*

2 Misologistic Nihilism

A second age-old source of philosophical discontent with the idea of human nature as morally normative comes in the form of misologistic nihilism. By misologistic nihilism I mean the despair of finding any meaning to life through the deliverances of human reason. Frustrated in various ways by reason, unable to discern a rationally secure moral frame of reference—indeed stymied by the confusing cacophony of arguments about *genesis* and *telos*, disputes about virtue and vice, and debates about human nature and happiness—misologistic nihilists abandon hope in reason and settle for skepticism, cynicism, or some kind of self-authorized meaning-making of their own. Misologistic nihilism is among the perils at hand when Socrates memorably warns Phaedo about the "danger of becoming misologists or haters of reason ... as people become misanthropists or haters of man, for no worse evil can happen to a man than to hate reason" (1914, p. 309, 89d). Misology comes about when "a man without proper knowledge concerning arguments has confidence in the truth of an argument and afterwards thinks that it is false, whether it really is so or not, and this happens again and again" (p. 313, 90b). It would be a "sad thing," he concludes, if one should end "by throwing the blame gladly upon the arguments and should hate and revile them all the rest of his life, and be deprived of the truth and knowledge of reality" (p. 313, 90d). The late Pope John Paul II had misologistic nihilism in his sights when he decried tendencies to reject "the meaningfulness of being":

> Quite apart from the fact that it conflicts with the demands and the content of the word of God, nihilism is a denial of the humanity and of the very identity of the human being. It should never be forgotten that the neglect of being inevitably leads to losing touch with objective truth and therefore with the very ground of human dignity. This in turn makes it possible to erase from the countenance of man and woman the marks of their likeness to God, and thus to lead them little by little either to a destructive will to power or to a solitude without hope (1998, §90).

Misologistic nihilists simultaneously deny what we are and despair of any rational resolution to the problem of human nature, thereby compounding one error with another one of far more deleterious consequence.

2.1 Misologistic Nihilism: Boethius's Burden and Reason's Recovery

Early in the *Consolation*, Lady Philosophy discerns in Boethius the dark despair emblematic of misologistic nihilism. In her first poem, she mourns Boethius's desolate frame of mind in vivid fashion:

> So sinks the mind in deep despair
> And sight grows dim; when storms of life
> Inflate the weight of earthly care,

> The mind forgets its inward light
> And turns in trust to the dark without.
> . . . Now see that mind that searched and made
> All Nature's hidden secrets clear
> Lie prostrate prisoner of night.
> His neck bends low in shackles thrust,
> And he is forced beneath the weight
> To contemplate—the lowly dust.

Burdened by manifest trouble and confused by goodness overrun by evil, Boethius despairs of any confidence in the "inward light" of the mind's eye. Absent trust in reason, he neglects the truths he once possessed, "turns in trust to the dark without," and neither contemplates the glories of heaven and earth, nor the profundities of life and death, nor Nature's divine master. His attention rather is turned to thinking about "the lowly dust"—which is to say, to thinking about not much at all.[17] Boethius confronts temptation, then, not only to scientific rationalism but also to misologistic nihilism.[18]

Lady Philosophy later draws the net more tightly around the mutually implicative relation between misology and nihilism. In a bit of perspicuous insight, while discussing the problem of evil, she queries Boethius at 4.pr.2:

> For I ask you, what is the cause of this flight from virtue to vice? If you say it is because they do not know what is good, I shall ask what greater weakness is there than the blindness of ignorance. And if you say that they know what they ought to seek for, but pleasure sends them chasing off the wrong way, this way too, they are weak through lack of self control because they cannot resist vice. And if you say they abandon goodness and turn to vice knowingly and willingly, this way they not only cease to be powerful, but cease to be at all. Men who give up the common goal of all things that exist, thereby cease to exist themselves. . . . A thing exists when it keeps its proper place and preserves its own nature. Anything which departs from this ceases to exist, because its existence depends on the preservation of its nature.

These lines offer much worthy of careful reflection. To be clear, Lady Philosophy is arguing that evil men are not powerful, as they seem to be, but rather are weak, and as such are unenviable. In so arguing, she presents three plausible explanations of "flight from virtue to vice," all of which suggest weakness on the part of those who are evil. Some opt for vice instead of virtue, in the first place, because of ignorance of what is good. They simply do not know what is good, and hence are not powerful but are weak. Though Francis Bacon did not elegantly assert that "knowledge is power" until much later, clearly Boethius embraces the same principle, and thus regards the failure to know as a weakness.[19] Others belong to a second category: they know what is good but run after vice instead of virtue because their will is unable to resist the allure of pleasure. As those who lack control over their very selves suffer the most fundamental form of frailty, Lady Philosophy relegates them, too, to the class of those who are not powerful but who to the contrary are weak. Finally, some belong to a third category of evil. Remarkably, these people know what is good, and they have self-control sufficient to choose in favor of what is good if they want it. Instead of knowing and willing what is good, however, they stunningly "abandon goodness and . . . not only cease to be powerful, but cease to be at all." If anything marks off weakness, a propensity toward self-annihilation would seem to do it, for whatever ceases to exist not only ceases to act, but clearly ceases to have the potency to act as well. And thus, Lady Philosophy concludes, for whatever reason evil men are vicious, we ought not to revere them because of any pretense of power that they claim.

Few things initially startle us more than the claim that knowledgeably and willfully choosing evil causes one to "cease to be." In what sense do such individuals cease to be? What does Lady Philosophy mean in alleging that these kinds of persons cease to be, and what assumptions underwrite her contention? After all, such individuals certainly do not appear to pass out of existence, for we find them within our experience of evil in the world. Indeed, the existence of resolutely evil people comprises the *sine qua non* of Boethius's grievous lament; if all evil was the result of simple ignorance or loss of self-control, he might have embraced Socrates' counsel that exhortative instruction rather than punishment is the appropriate response to wrongdoing (see, e.g., *Apology* 26a). He does not do so; he instead complains bitterly about the existence of willfully evil people, people who deserve and should receive punishment in a providentially ordered world, but who, alas, do not. What is more, Lady Philosophy takes his complaint seriously, and she offers a stark assessment: those who willfully embrace evil, and do so understanding what they do, "cease to be at all."

Lady Philosophy's astonishing claim at a minimum means this: human beings who knowledgeably and willfully choose evil rather than good act contrary to their nature. In so acting, they undermine the very principle that makes them what they are. Acting against their nature carries a grave consequence: they invariably become less than what they properly are, or worse. As Lady Philosophy explains, it is the nature of human beings to be rational and mortal animals that share in the "common goal of all things that exist," namely goodness.[20] Because human beings are not merely animals, but are rational animals, human beings do not merely share in the "common goal of all things that exist," but share in the "common goal of all things that exist" in a way commensurate with their nature, which is to say that human beings share in goodness rationally. That is what it means to be a human being; that is what human nature is; and that, I should say, is what makes human nature morally normative on Boethius's account. Moreover, that is what Lady Philosophy means when she explains, "Men who give up the common goal of all things that exist, thereby cease to exist themselves. . . . Anything which departs from this ceases to exist, because its existence depends on the preservation of its nature" (4.pr.2).

In the next section of book four, Lady Philosophy elaborates more vividly. Arguing that the "punishment of the wicked is their very wickedness," she explains:

> . . . anything which turns away from goodness ceases to exist, and thus . . . the wicked cease to be what they once were. That they used to be human is shown by the human appearance of their body which still remains. So it was by falling into wickedness that they also lost their human nature. Now, since only goodness can raise a man above the level of human kind, it follows that it is proper that wickedness thrusts down to a level below mankind

> those whom it has dethroned from the condition of being human. … So what happens is that when a man abandons goodness and ceases to be human, being unable to rise to a divine condition, he sinks to the level of being an animal (4.pr.3).

The result, Lady Philosophy says, is that "you cannot think of anyone as human whom you see transformed by wickedness" (4.pr.3). To lose or give up the rational disposition to goodness—to participation in the "common goal of all things that exist"—is to lose or give up one's "true self."

Perhaps it is useful to put the point positively. Over the long course of the *Consolation*, Lady Philosophy brings Boethius to see that we are rational and mortal animals whose beginning and end lies in divine goodness. Quite simply, it is the nature of human beings to be rationally disposed to goodness; rationally apprehending and choosing what is good fulfills human nature, and any failure to rationally apprehend and choose what is good frustrates human nature. Human nature is therefore morally normative in the straightforward sense that knowing what one is sheds light upon what one ought and ought not to do. The virtues, many of which Boethius embodies, represent varied ways of rationally apprehending and choosing good. Prudence, justice, fortitude, moderation, and all of the other virtues are virtues just and only because they help realize or fulfill human nature so defined. The habitually prudent person reasons well about what is good; the habitually moderate person consistently chooses what is good in proper rather than excessive or deficient proportions; and so on. Viciousness, of whatever form, is what it is because it undermines or frustrates human nature so defined. The habitually imprudent person fails to reason well about what is good; the habitually immoderate person consistently desires too little or too much of what is good, even if they know what is good; and the habitually unjust person, at least in the worst case, inexplicably chooses evil in spite of knowing what is good. In all of these cases, viciousness makes one less than fully human; it undermines one's rational orientation to goodness. It should be said that Boethius's experience of loss and lament, and his despairing forgetfulness of what he is, signals a degradation of both his mind and his will, and it precipitates Lady Philosophy's two-fold therapeutic regimen: philosophical argument to restore proper understanding within his addled mind, and sweet poetry to rekindle properly ordered love within his disarrayed will.

2.2 Camus, Sartre, and Percy: Contemporary Heralds of Misologistic Nihilism

We do not have far to look in order to find contemporary forms of misologistic nihilism. Albert Camus is one *enfant terrible* among others. His outlook—at least in some of his writing, especially of earlier vintage—is so severe as to have won the appellation of absurdism. According to Camus in *The Myth of Sisyphus*, there is no such thing as human nature or a human essence, and consequently all attempts to find a rational meaning for human existence are doomed. Indeed, he provides a self-reflective account of his own experiences, as one among other dispossessed denizens of late modernity, that perfectly fits the aetiological profile Socrates offers for the development of misology. Camus writes:

> We must despair of ever reconstructing the familiar, calm surface which would give us peace of heart. After so many centuries of inquiries, so many abdications among thinkers, we are well aware that this is true for all our knowledge. With the exception of professional rationalists, today people despair of true knowledge. If the only significant history of human thought were to be written, it would have to be the history of its successive regrets and its impotences (1991, p. 18).

Held in despairing captivity to the "successive regrets" and "impotences" of human thought, Camus can claim no more than that: "This heart within me I can feel, and I judge that it exists. This world I can touch, and I likewise judge that it exists. There ends all my knowledge, and the rest is construction" (p. 19). Given that reason's resources are insufficient to make good on the existential debits life presses upon us—particularly in the form of the conspicuous evils of a world imperiled by mid-twentieth-century totalitarianism—he concludes that there is no meaning to be found in the "absurd abyss of existence," and if we are honest about the matter, we will realize that "There is but one truly serious philosophical problem, and that is suicide" (p. 3). In the face of our plight, so it would seem on Camus's view, we can deceive ourselves, we can destroy ourselves, or we can drive deliberately onward through life, living as absurd heroes in spite of the abyss.[21]

Camus's one-time friend, Jean-Paul Sartre, embraces a similarly skeptical stance toward the possibility of anything like a "human nature." Because Sartre does not think that human beings have a nature, the notion of a morally normative human nature is also for him, *ipso facto*, wrongheaded at the most metaphysically basic of levels. As he puts it:

> Atheistic existentialism, of which I am a representative, declares [that] . . . there is no human nature, because there is no God to have a conception of it. Man simply is. . . . Man is nothing else but that which he makes of himself. That is the first principle of existentialism (1948, pp. 27–28).[22]

Like Boethius prior to the artistic and intellectual salves plied by Lady Philosophy, Sartrean existentialism forgets—or neglects or desists from—what we are. It does so in the most straightforward of ways, by denying that human beings have a nature at all, and doing this as the natural corollary of denying that of which Boethius too, at least momentarily, loses sight, viz. humanity's source and end in God. The question of whether or not human nature is morally normative thus has no purchase for Sartre, since it assumes something in doubt. To the contrary, "Man makes himself; he is not found ready-made; he makes himself by the choice of his morality, and he cannot but choose a morality, such is the pressure of circumstances upon him" (1948, p. 50).[23] Sartre's propensity for misologistic nihilism is if anything starker than Camus's.

Yet however explicitly Camus and Sartre embody late modernity's dubiousness about anything like "knowledge" of what we are, I nonetheless want to attend to a third writer of the twentieth century, Walker Percy, in whose literary *opus* misologistic nihilism is rife. Percy, of course, is intent upon trenchant and even scathing critique of the desiccated culture of modernity that can make such ideas *au courant*. Thus is it that we encounter among the fascinating *dramatis personae* of Percy's *Moviegoer* the remarkable Binx Bolling, a young man who looks intently into the conditions of misologistic nihilism in our day. Malaise, everydayness, repetition, despair—such terms of the trade take form on the pages of Percy's book as he makes out Binx's life, one lived, he self-consciously narrates, in

> the very century of *merde*, the great shithouse of scientific humanism where needs are satisfied, everyone becomes an anyone, a warm and creative person, and prospers like a dung beetle, and one hundred percent of people are humanists and ninety-eight percent believe in God, and men are dead, dead, dead; and the malaise has settled like a fall-out and what people really fear is not that the bomb will fall but that the bomb will not fall (1998, p. 228).

Ralph Wood remarks that Percy's "fiction makes a withering critique of what is spiritually inane about contemporary American life" (1988, p. 133). In this respect, Percy shows himself to be a latter-day student of Lady Philosophy, for her first order of business in the *Consolation*, before uttering a word to Boethius, is to express anger at the Muses of Poetry who bring false "comfort in hapless age" (1.m.1). Philosophy condemns them as "hysterical sluts" with "no medicine to ease his pains, only sweetened poisons to make them worse," the "very creatures who slay the rich and fruitful harvest of Reason with the barren thorns of Passion." Because they "habituate men to their sickness of mind instead of curing them," Philosophy drives away these damnable Sirens and insists that they "leave him for my own Muses to heal and cure" (1.pr.1). Percy is no less intent to pronounce anathema upon any infatuation with sham therapies that distract and disorient us from genuine self-understanding. Wood explains that because Percy "refuses to tolerate the malodorous character of our common life," he aims in his fiction to identify "the public and private stench for the foul thing that it is" (1988, p. 133).

Percy's indebtedness to Camus is widely acknowledged.[24] For this reason it is unsurprising that *The Moviegoer* treats the scope of Camus's taxonomy of life options: self-deception, self-destruction, and absurd heroism. Binx recognizes that most people do not like to think, and thus they inoculate themselves against reason by uncritically adopting culturally sanctioned ideas about life's meaning and purpose. Because he believes that the culturally ascendant ideas in vogue are existentially vacuous (e.g., consumerism, hedonism, nationalism), those who endorse them tend to be self-deceiving, despairing nihilists, whether or not they know it.[25] Binx also confronts the bleak attraction of self-destruction. Though perhaps contemplated inchoately in his own mind, the question of suicide is demonstrably a live one in Kate Cutrer, to whom Binx becomes married by the end of the novel. Kate struggles with suicidal impulses, repeatedly acquiring and casting away bottles of sodium pentobarbital, living in fear of "a general catastrophe" (1998, p. 28).[26] Speaking of Percy's abiding attention to the question of suicide, Wood observes that "though often a merely selfish act, the taking of one's own life can also serve as a way of asking whether human existence is ultimately worth affirming" (1988, pp. 134–135). Far from a mental defect, from one vantage point the temptation to self-destruction may constitute a perfectly rational act in an apparently irrational world. Wood notes:

> It raises the metaphysical quandary that ethics alone cannot answer. Ethics determines why we should do one thing rather than another: why to act with noble rather than base motives, why to live cowardly or courageous lives, why to speak truthfully rather than deceptively. For Percy the real question is not ethical but metaphysical and theological: why should one do *anything* at all? What is there, either within human life or beyond it, that justifies all of its trouble? This most fundamental of problems Percy's fiction relentlessly seeks to analyze and also to answer (1988, p. 135).

The question of suicide in Percy's fiction thus gives lucid expression to misologistic nihilism in its most life-wrenching form. In the absence of cogent answers to the basic questions of human existence, why should one continue living? Discomfiture about reliable solutions to the riddle of human nature easily gives way to a self-destructive despair, unless it resolves in the dishonest direction of self-deception or else, on Camus's account, in the nobility of absurd heroism. Percy toys with the latter possibility—for Camus paradigmatically represented by the mythical Sisyphus—in Binx's fascination with "successful repetition"[27] and "good rotation."[28] However one gets about the business of discontent with human nature, the end result, at least for Binx, is despair. Whether through self-deceptive rationalization, self-annihilation, or absurd slogging onward in spite of the meaninglessness of it all, misologistic nihilism looms large on the landscape of the twentieth-century American culture in which Percy locates Binx.

Yet after all of this, Binx himself falls outside the scope of Camus's taxonomy of life's options: he does not settle for the self-deceptive forms of coping which typify the masses; though perhaps tempted to suicide, he does not kill himself; and he does not nobly endure the absurdity of it all. He is—though somewhat tentative about it—a seeker.

> What is the nature of the search? you ask.
>
> Really it is very simple, at least for a fellow like me; so simple that it is easily overlooked.
>
> The search is what anyone would undertake if he were not sunk in the everydayness of his own life. This morning, for example, I felt as if I had come to myself on a strange island. And what does such a castaway do? Why, he pokes around the neighborhood and he doesn't miss a trick.
>
> To become aware of the possibility of the search is to be onto something. Not to be onto something is to be in despair (1998, p. 13).

Unlike the Camus of *The Myth of Sisyphus*, Percy's Binx Bolling sees a *tertium quid* other than absurd heroism: the search. Without knowing in advance the fullness of what he will find, Binx discerns the "possibility of the search," finds himself "onto something," and thus avoids the despairing malaise of the masses' everydayness.

In this remarkable respect, Percy puts Binx squarely in the same frame of reference within which Lady Philosophy attempts to reorient Boethius. Binx could not be more different in this sense from Camus and Sartre; in response to both of these heralds of misologistic nihilism, Lady Philosophy surely would confront a predicament. The things with which she can work in treating Boethius's moral malaise—his acknowledgement of God as Creator, the world as meaningfully ordered, and humankind as by nature rational and mortal—these are the very things they deny. But Binx is different. Though disgusted by the sham solutions of spiritualists, Christian or otherwise, Percy acknowledges that "there is hardly a moment in my writing when" the protagonist of his novel, "usually some kind of Catholic, bad, half-baked, lapsed, whatever . . . stands vis-à-vis the Catholic faith" (Quinlan, 1996, p. 1). This general rule of Percy's art undeniably captures the dramatic trajectory of *The Moviegoer*, the main part of which concludes perhaps too obviously between a school and a church on Ash Wednesday, with Binx speculating about a penitent's ambiguous purposes and wondering whether "through some dim dazzling trick of grace" he encountered the divine as "God's own importunate bonus" (1998, p. 235). Describing his purposes in *The Moviegoer*, Percy later made the point prosaically: "the book attempts a modest restatement of the Judeo-Christian notion that man is more than an organism in an environment, more than an integrated personality, more even than a mature and creative individual, as the phrase goes. He is a wayfarer and a pilgrim" (1991, p. 246). He is a rational and mortal animal whose beginning and end lie in divine goodness. He is *homo viator*, whose lifelong quest, Percy writes, must be to figure out the significance of "signposts in a strange land" and to look out for "the message in a bottle" with its piece-of-news-from-across-the-seas importance.

Though Percy stands with Camus and Sartre as a contemporary herald of misologistic nihilism, Percy's work unmistakably deserves our utmost care in acknowledging its "hints at a way beyond our current malaise" (Wood, 1988, p. 133). Percy is no more willing to see late-modern humanity deprived of its rational and moral dignity than Lady Philosophy is prepared to see her own downtrodden devotee in despair. He thus works strenuously—as Camus and Sartre do not—to retrieve, reclaim, and renew a morally normative conception of human nature. Moreover, like Boethius's wise and artfully adept mistress, Percy understands that our despairing neglect of what human nature is and what it portends signals the ruinous condition, in our late-modern culture, of both the human mind and will. Just as Lady Philosophy makes use of argument to puts Boethius's mind in order and beautifully-wrought verse to reorient his will, so also, therefore, does Percy's *oeuvre* employ the same techniques. In some cases, such as in *The Message in the Bottle* or in *Signposts in a Strange Land*, arguments come to the fore. In other instances, such as in *The Moviegoer* or *Love in the Ruins*, Percy's literary craftsmanship more indirectly but nonetheless efficaciously guides his readers to the same conclusions. In still other instances, Percy intimates the most direct and forceful of possible remedies to late-modern confusion about human meaning, nature, and purpose: what we really need, he suggests, is a "good kick in the ass." We find such a candid response to the neglect of obvious features of human nature, for instance, in *The Moviegoer* when Binx pillories the insipidity of the long-running series "This I Believe" (recently revived by National Public Radio—unbelievably—for the twenty-first century). After commenting upon the claptrap typical of "This I Believe," Binx reports once recording "a tape which I submitted to Mr Edward R. Murrow. 'Here are the beliefs of John Bickerson Bolling, a moviegoer living in New Orleans,' it began, and ended,

'I believe in a good kick in the ass. This—I believe'" (1998, p. 109). Boethius, so it would seem, similarly intuited that the rigors of logic and the loveliness of poetry can fall short of full effect unless preceded by aggressive rousing from self-preoccupation. After all, he occasionally finds his own shrewd mistress prepared to address him forthrightly and importunately, as when she protests, "I can't put up with your dilly-dallying and the dramatization of your care-worn grief-stricken complaints that something is lacking from your happiness" (2.pr.4). This, *inter alia*, constitutes Boethius's sixth-century version of Binx Bolling's good, swift kick in the backside. Along with both Boethius's Lady Philosophy and Percy's Binx Bolling, perhaps we too on this count should intone, "This—I believe."

I have tried to make out two ways in which we, at the end of modernity, have forgotten what we are. In following Lady Philosophy's efforts to move Boethius from desolation to consolation, I have argued that scientific rationalism and misologistic nihilism represent basic forms of getting human nature wrong. In both cases, human beings are wrongly regarded as something less or other than rational and mortal animals whose beginning and end lies in divine goodness, and in both cases human beings are consequently left without rationally adequate and morally significant means for being what in truth they are. Moreover, I have attempted to show that far from the latest, faddish instances of intellectual *haute couture*, scientific rationalism and misologistic nihilism are perennial and persistent forms of "getting it wrong" when it comes to the question of human nature and its moral significance. These two potential errors confront Boethius in his real experience, and they confront the character bearing his name and voicing his laments in *The Consolation of Philosophy*. Our own contemporary culture is, to be sure, marked by widespread expressions of scientific rationalism and misologistic nihilism, demonstrably in the first instance by the likes of Daniel Dennett and in the second instance by Albert Camus, Jean-Paul Sartre, and company. Not least of all, I have intimated how it is that philosophical knowledge of what human beings are provides morally normative direction for our lives, potentially consoling and curing us in the same way that Lady Philosophy heals Boethius's mind and will. In connection with this last claim, finally, I have singled out Walker Percy not only as a contemporary herald of misologistic nihilism, but more importantly as a herald—like Boethius of old—of the consolation possible beyond scientific rationalism and misologistic nihilism.

Here at the end, however, I must offer a caveat, one that follows Lady Philosophy all the way to the conclusion of Boethius's *Consolation*. In short, there are no guarantees that true philosophy or good poetry, either singly or jointly, can put right muddled minds or wrecked wills. Precisely on account of this, Lady Philosophy points beyond her own powers in the closing lines of the *Consolation*. She exhorts Boethius:

> Hope is not placed in God in vain and prayers are not made in vain, for if they are the right kind they cannot but be efficacious. Avoid vice, therefore, and cultivate virtue; lift up your mind to the right kind of hope, and put forth humble prayers on high. A great necessity is laid upon you, if you will be honest with yourself, a great necessity to be good, since you live in the sight of a judge who sees all things (5.pr.6).

Lady Philosophy reminds Boethius that remembering what we are, with all its implications for what we ought and ought not to do, requires remembering whose we are, with all of the portentous implications concomitant to such weighty self-understanding. When all is said and done, then, the question of whether or not human nature is morally normative, and in what ways, may depend entirely upon whether one does or does not find it possible to embrace a faith beyond reason.

Notes

1. Unless otherwise noted, quotations of this text will be from Boethius (1999), and will be parenthetically noted by book, by prosa or metra section number, and occasionally by line number. Latin citations will be from James J. O'Donnell's text of the *Consolatio Philosophiae*, derived from G. Weinberger (1935).
2. Hugh Fraser Stewart's fine treatment, *Boethius: An Essay* (1974; reprint, 1891), maintains that Boethius comes to accept this as his own judgment about himself: "He shall lay bare his inmost heart to her, and confess that indeed he knows not what he is, nor what man himself is" (p. 59).
3. Regarding "the importance of his books in medieval education," Howard Rollin Patch simply says, "The enormous number of manuscripts and the fact that they were widely used in the universities tell their own story" and follows up by asking "How then may one even approximate an accurate analysis of the influence that Boethius exerted on his fellow men, when his offering was so rich and so varied?" (1935, p. 116). Patch also writes, "Seldom has any author pervaded so thoroughly, and even formed, the thought and the expression of his own and later periods" (p. 6). C. S. Lewis says of Boethius's *De Consolatione Philosophiae* that it "was for centuries one of the most influential books ever written in Latin. It was translated into Old High German, Italian, Spanish, and Greek; into French by Jean de Meung; into English by Alfred, Chaucer, Elizabeth I, and others. Until about two hundred years ago it would, I think, have been hard to find an educated man in any European country who did not love it" (1964, p. 75). In response to a slight of Boethius by Edward Gibbon, he adds, "It is historically certain that for more than a thousand years many minds, not contemptible, found it nourishing" (p. 90). And Henry Chadwick says that by "writing the *Consolation of Philosophy* Boethius provided all educated people of the Middle Ages and the Renaissance with one of their principal classics" (1981, p. 1.)
4. I grant that to speak of "scientific rationalism" and "misologistic nihilism" is to identify the issues in terms more familiar to us than to Boethius. Yet while perhaps no cultural milieu has given as popular and forceful an expression to scientific rationalism and misologistic nihilism than that of late modernity, neither tendency is unique to our time. To the contrary, they have ancestral forms that considerably antedate Boethius's own sixth-century Roman context, although they took on forms familiar to Boethius within his own era. That late modernity finds in these two age-old frames of reference a ready basis for discontent with what we are simply underscores the philosophically predictable, if pernicious, forms of error.
5. Alas, a recent pretender to Boethius's great work—at least as suggested by title of Alain De Botton's *The Consolations of Philosophy* (2000)—prescinds from all of the straightforwardly normative judgments identified here that are so well evidenced in Boethius's *Consolatio Philosophiae*.
6. Post goes on to explain that a "special case of naturalism restricts the basic entities to those that mathematical physics investigates (such as the various subatomic particles, force-fields, and variably curved space-time). In this version of naturalism, called 'physicalist materialism,' or 'physicalism,' everything is some collection of the basic entities that mathematical physics studies, and all truth is determined ultimately by the truths about these entities" (1991, p. 11).
7. In addition to Post's *Metaphysics: A Contemporary Introduction*, see John F. Post (1987). Daniel Dennett, discussed below, also tries to blunt the edge of reductionism in "Who's Afraid of Reductionism?" (1995, pp. 80–83).

8. David Hull writes that "Dennett's views are reductionism incarnate, a conclusion that Dennett himself explicitly avows, but he also distinguishes between two sorts of reductionism—good and greedy. In its most general form, good reductionism is merely the rejection of any supernatural explanations of natural phenomena. ... According to the greedy version of reductionism, scientists should abandon all the principles, theories, vocabulary, and laws of the higher-level sciences for those of the lowest-level sciences, explaining, for example, the decisions of the Rehnquist Court in terms of entropy fluctuations. Dennett takes this version of reductionism to be preposterous" (1996, pp. 171–172).
9. I can regard Dennett more sympathetically than those the late Pope John Paul II must have had in mind when writing: "Another threat to be reckoned with is *scientism*. This is the philosophical notion which refuses to admit the validity of forms of knowledge other than those of the positive sciences; and it relegates religious, theological, ethical and aesthetic knowledge to the realm of mere fantasy. In the past, the same idea emerged in positivism and neo-positivism, which considered metaphysical statements to be meaningless. Critical epistemology has discredited such a claim, but now we see it revived in the new guise of scientism, which dismisses values as mere products of the emotions and rejects the notion of being in order to clear the way for pure and simple facticity. Science would thus be poised to dominate all aspects of human life through technological progress. ... Regrettably, it must be noted, scientism consigns all that has to do with the question of the meaning of life to the realm of the irrational or imaginary" (1998, Chapter 7, § 88).
10. Other fine efforts, of course, have been and are being made. Among the latest, for instance, see the work of William D. Casebeer (2003).
11. Similarly, he writes, "People ache to believe that we human beings are vastly different from all other species—and they are right! We are different. We are the only species that has an *extra* medium of design preservation and design communication: culture" (p. 338). Subsequently, he writes that "...there is a huge difference between our minds and the minds of other species, a gulf wide enough even to make a moral difference" (p. 371).
12. Chadwick's essay in Margaret Gibson's outstanding edited volume (published in commemoration of the fifteen hundredth anniversary of the birth of Boethius) makes the point that in "its philosophical content the *Consolation* attracted commentaries from several medieval authors, ... a significant sign of the seriousness with which men took his philosophical reflections on the dealings of providence with a world beset by so much evil" (p. 1). In fact, while a range of issues gave rise to longstanding efforts at commentary (including, for instance, the relative weight of pagan and Christian influences in the text), there is no more important key to a fruitful philosophical interpretation of the text than discerning the centrality of the problem of evil, addressed by Boethius in its existential, philosophical, and theological dimensions.
13. The substantive argument that brings Boethius to this conclusion begins in book three of the *Consolation* and extends into book four. It involves a series of claims about the rationally apprehensible and self-evidently desirable good of true happiness, along with an identification of God as "the essence of happiness" (3.pr.10). Based upon this identification, Lady Philosophy argues, "Since it is through the possession of happiness that people become happy, and since happiness is in fact divinity, it is clear that it is through the possession of divinity that they become happy. ... While only God is so by nature, as many as you like may become so by participation" (3.pr.10). She furthermore argues that all living things, self-consciously or not, desire goodness, beginning first and minimally with self-preservation and growing out from there to encompass greater qualities of goodness, and thus it is that "the end of all things is the good" (3.pr.11)—a good that in the case of human beings is rationally discerned, willingly desired, and morally normative in its orientation toward participation in the divine life. Because "God controls all things by the helm of goodness, and all things ... are willingly governed and willingly obey the desires of him who controls them" (3.pr.12), Lady Philosophy lastly insists that human beings can lodge confidence in their longing for happiness and in their exercise of reason on behalf of such happiness, notwithstanding their apparently deleterious encounters with evil.
14. Several reviewers of Dennett's work express extreme dubiousness about his claims. Thus, Gary Mar writes that Dennett "applies Darwinism to the human species—to the mysteries of conscious-

ness, language, mathematics, and morality. These mysteries are even more wonderful, says Dennett, when they turn out not to be 'sacred' but scientifically explicable using only evolutionary 'cranes' without any pseudo-explanatory 'skyhooks' (Dennett's metaphor for a *deus ex machina*)" (1996, p. 26). Alvin Plantinga follows Mar's own subsequent expression of suspicion about Dennett's triumphalism (1996). Plantinga points to the book's "pervasive tendentiousness, or perhaps a certain list toward demagoguery" (p. 16); its "great deal of unbridled speculation as well as much very energetic hand waving" about the naturally selective grounds for our moral proclivities (p. 17); and its failure to see real dangers with Darwinism, which Plantinga describes by explaining:

> . . .there is one respect in which Darwin's dangerous idea is vastly more dangerous than Dennett realizes. According to Richard Rorty,
>
>> The idea that one species of organism is, unlike all the others, oriented not just toward its own increased prosperity but toward Truth, is as un-Darwinian as the idea that every human being has a built-in moral compass—a conscience—that swings free of both social history and individual luck. . . .
>
> [Rorty] says that two ideas are un-Darwinian: that we have a mind oriented toward the Truth and a conscience that puts us in touch with right and wrong.

The problem with Dennett, Rorty might suggest, is a failure of irony. Dennett wants the benefits of a morally normative account of human nature without the metaphysical apparatus to sustain it. Unlike Rorty, who is content to abandon non-ironic talk of truth, conscience, morality, and the like, Dennett keeps up the appearance of straightforward talk about such matters. Plantinga, of course, rejects Rorty's position, though he admires its no-nonsense honesty. He plainly has Dennett in mind when he concludes, "Modern science was conceived, and born, and flourished in the matrix of Christian theism. Only liberal doses of self-deception and double-think, I believe, will permit it to flourish in the context of Darwinian naturalism" (p. 35).

Other reviewers' criticisms are equally trenchant. In his generally sympathetic treatment, David Hull tellingly writes that Dennett has "not much" to say about "the subtitle of the book—evolution and the meanings of life" (p. 173). Another reviewer notes that what "counts as good Dennett never directly addresses, which leaves the account problematic," and points to "serious flaws in Dennett's central position and treatment of religious viewpoints" (Peterson, 1998, pp. 160–161). Yet another reviewer complains that "when Dennett looks at moral values in a Darwinian context, the results are not earth-shaking. . . . Dennett is convinced that only evolutionary analysis can make sense of the origins of ethical norms, but in the final analysis, he sheds little new light on the key question of how selective pressure might produce characteristics of altruism and cooperation" (Hester, 1997, p. 247). And another critic writes that Dennett "does not really fulfill the promise to put life's meanings on a new foundation, but he is in the familiar predicament of the disputant who lacks any metaphysics" (Gallagher, 1996, p. 488).

15. In an especially scathing review, evolutionary geneticist Allen Orr makes out the dearth of evidential support for the conclusions Dennett draws, and by implication the improbability of forthcoming data on behalf of Dennett's hyperbolic conjectures:

> The deeper point, though, is that this navel-gazing—these endless attempts to theoretically reconstruct what 'must have' occurred during the emergence of human morality—is no more than academic exercise. The ugly fact is that we haven't a shred of evidence that morality in humans did or did not evolve by natural selection. We do not even know what such evidence would look like. We can, if we like, construct plausible adaptive scenarios ('What would happen to a gene that said be nice to strangers if. . .'). But, in the end, a thought experiment is not an experiment. We have no data.
>
> Dennett's treatment of evolutionary ethics is symptomatic of the problem plaguing his entire book. He is forever suggesting that the universal acid of natural selection *may* be involved here or there. . . . But at each milepost the skeptical reader grumbles, 'But maybe not.' After all, the evidence for each claim ranges from non-existent (alternative universes, origin of morality) to negative (Darwinian evolution of memes). All Dennett really shows is that—if one squints hard enough—one can *sort of* see how Darwin's dangerous idea *might* play a role in this,

that, or the other. Although he has produced a provocative and intermittently entertaining book, Dennett's chief aim is unconvincing. Darwinism may have little to tell us outside of biology (Orr, 1996, p. 471).

Dennett responds to a version of Orr's critique reprinted in the *Boston Review*. See Dennett (1996).

16. Later in the same essay he writes: "God is absent, said Johann Christian Hölderlin; God is dead, said Nietzsche. This means one of two things. Either we have outgrown monotheism, and good riddance; or modern man is estranged from being, from his own being, from the being of other creatures in the world, from transcendent being. He has lost something—what, he does not know; he knows only that he is sick unto death with the loss of it" (p. 262). Percy's experiences and observations resonate strongly with those of Boethius as Lady Philosophy finds him at the outset of the *Consolation*.
17. "Stolidam cernere terram," the poem concludes. Literally, it would seem, one of the sort here in question ends up sifting the dull dirt, about the most meaningless and unfruitful activity imaginable.
18. John Paul II astutely observes that nihilism is often experientially prompted by our encounter of evil, and also that where nihilism does not set in, scientific rationalism often takes its place. "This nihilism has been justified in a sense by the terrible experience of evil which has marked our age. Such a dramatic experience has ensured the collapse of rationalist optimism, which viewed history as the triumphant progress of reason, the source of all happiness and freedom; and now, at the end of this century, one of our greatest threats is the temptation to despair. Even so, it remains true that a certain positivist cast of mind continues to nurture the illusion that, thanks to scientific and technical progress, man and woman may live as a demiurge, single-handedly and completely taking charge of their destiny" (1998, Chapter 7, § 91).
19. Francis Bacon's claim that "ipsa scientia potestas est" appears in his *Meditationes Sacræ. De Hæresibus* (1597) (*Religious Meditations, of Heresies*).
20. Cf. *Consolation* 4.pr.2.11–12: "num recordaris beatitudinem ipsum esse bonum eoque modo, cum beatitudo petitur, ab omnibus desiderari bonum? – minime, inquam, recordor, quoniam id memoriae fixum teneo. omnes igitur homines boni pariter ac mali indiscreta intentione ad bonum peruenire nituntur?—ita, inquam, consequens est."
21. I acknowledge the judgment of some that Camus's later fiction expresses yet another possibility, namely the attractiveness of something like the *anima naturaliter Christiana*—of which fellow French novelist François Mauriac accused Camus himself. See Lottman's *Albert Camus: A Biography* (1979, p. 452). On the same subject, Lottman notes that following the appearance of *La Chute* (*The Fall*) in 1956, critics were surprised to discover in the book "Christianity (the hero's name is of course an allusion to John the Baptist, 'the voice of one crying in the wilderness') and Dostoevsky as well. When Conor Cruise O'Brien pointed out in *The Spectator* that the novel had Christian overtones, Camus confirmed, in a letter to his British publisher, that this approach was justified" (p. 579). Lottman relates an account of Camus's life over the course of which he "would be invited, cajoled, into a commitment on Christianity, which he never made," and yet a life fraught with sufficient equivocation toward transcendent horizons of meaning that, "After his death, believers often left crosses on his grave (which is otherwise singularly undecorated)" (p. 290). Lottman also writes that "Communist philosophers called him reactionary, reactionaries said he was a Communist, atheists that he was a Christian, while Christians deplored his atheism" (p. 615). Lottman seems to split the difference and thus to sustain uncertainty in noting, "Essays and even books have been written about a Christian Camus, and they could not have been difficult to write, for Camus seeded the ambiguities" (p. 290). Olivier Todd's more recent work in *Albert Camus: A Life* (2000; reprint 1997) attempts to address some of these same issues. He relates Camus's claim in a December 1946 lecture to be "your Augustine before his conversion. I am debating the problem of evil, and I am not getting past it" (p. 230). Still, he pulls no punches in acknowledging that "Christians took note of Camus's atheism but expected him to convert one day, when in fact he never had any faith" (p. 230).
22. Sartre unambiguously rejects the view that "Man possesses a human nature; that 'human nature,' which is the conception of human being, is found in every man; which means that each man is a particular example of an universal conception, the conception of Man" (p. 27).

23. Philip Mairet elaborates, "Although his personal fate is simply to perish, [man] can triumph over it by inventing 'purposes,' 'projects,' which will themselves confer meaning both upon himself and upon the world of objects—all meaningless otherwise and in themselves. There is indeed no reason why a man should do this, and he gets nothing by it except the authentic knowledge that he exists; but that precisely is his great, his transcendent need and desire. Not many are capable thus of authenticating their existence: the great majority reassure themselves by thinking as little as possible of their approaching deaths and by worshipping idols such as humanity, science, or some objective divinity" (1948, pp. 14–15).
24. In one place Percy writes: "The hero of the postmodern novel is a man who has forgotten his bad memories and conquered his present ills and who finds himself in the victorious secular city. His only problem now is to keep from blowing his brains out" (1975, p. 112). For more on Percy's indebtedness to Camus, see Wood (1988, pp. 134–135); Quinlan (1996, p. 17); and Rudnicki (1999, p. 34).
25. The epigraph to Percy's book, taken from Søren Kierkegaard's *Sickness Unto Death*, must be borne in mind: "...the specific character of despair is precisely this: it is unaware of being despair."
26. Percy's lifelong struggle to come to terms with repeated suicides within his family lineage, including his own father's, is well known.
27. Percy has Binx explain: "What is a repetition? A repetition is the re-enactment of past experience toward the end of isolating the time segment which has lapsed in order that it, the lapsed time, can be savored of itself and without the usual adulteration of events that clog time like peanuts in brittle. Last week, for example, I experienced an accidental repetition. I picked up a German-language weekly in the library. In it I noticed an advertisement for Nivea Creme, showing a woman with a grainy face turned up to the sun. Then I remembered that twenty years ago I saw the same advertisement in a magazine on my father's desk, the same woman, the same grainy face, the same Nivea Creme. The events of the intervening twenty years were neutralized, the thirty million deaths, the countless torturings, uprootings and wanderings to and fro. Nothing of consequence could have happened because Nivea Creme was exactly as it was before" (1998, pp. 79–80).
28. According to Binx, "A rotation I define as the experiencing of the new beyond the expectation of the experiencing of the new" (1998, p. 144).

Bibliography

Boethius (1999). *The consolation of philosophy*, rev. ed., V. Watts (trans.). New York: Penguin.

Camus, A. (1991). *The myth of Sisyphus and other essays*, J. O'Brien (trans.). New York: Vintage International.

Casebeer, W.D. (2003). *Natural ethical facts: Evolution, connectionism, and moral cognition*. Cambridge: MIT Press.

Chadwick, H. (1981). Introduction in *Boethius: His life, thought, and influence*, M. Gibson (ed.). Oxford: Basil Blackwell.

De Botton, A. (2000). *The consolations of philosophy*. New York: Vintage.

Dennett, D. (1995). *Darwin's dangerous idea: Evolution and the meanings of life*. New York: Simon and Schuster.

Dennett, D. (1996). The scope of natural selection. *Boston review: A political and literary forum*, vol. 21, no. 5 (October/November). [On-line.] Available: http://www.bostonreview.net/br21.5/dennett.html. (Accessed March 09, 2009).

Gallagher, K.T. (1996). Review of *Darwin's dangerous idea*, by Daniel C. Dennett. *International philosophical quarterly* 36 (4):487–489.

Hester, C. (1997). Review of *Darwin's dangerous idea*, by Daniel C. Dennett. *Journal of scientific exploration* 11 (2): 243–249.

Hull, D.L. (1996). Review of *Darwin's dangerous idea*, by Daniel C. Dennett. *Ethics* 107 (1): 170–174.

Lewis, C.S. (1964). *The discarded image: An introduction to medieval and renaissance literature*. Cambridge: Cambridge University Press.
Lottman, H.R. (1979). *Albert Camus: A biography*. Garden City, New York: Doubleday.
Mairet, P. (1948). Introduction, in *Existentialism and humanism* by Jean-Paul Sartre. London: Methuen.
Mar, G. (1996). Review of *Darwin's dangerous idea*, by Daniel C. Dennett. *New oxford review* 63(6) (July-August).
O'Donnell, J.J. (1935). *Consolatio philosophiae*, vol. 67, derived from G. Weinberger. Vienna: Corpus Scriptorum Ecclesiasticorum Latinorum.
Orr, H.A. (1996). Dennett's dangerous idea, review of *Darwin's dangerous idea*, by Daniel C. Dennett, *Evolution* 50 (1): 467–472.
Patch, H. R. (1935). *The tradition of Boethius*. New York: Oxford University Press.
Paulus, I. PP. II (1998). *Fides et ratio*. Vatican City: Libreria Editrice Vaticana.
Peterson, G.R. (1998). Review of *Darwin's dangerous idea*, by Daniel C. Dennett. *Zygon* 33 (1): 158–161.
Plantinga, A. (1996). Review of *Darwin's dangerous idea*, by Daniel C. Dennett. *Books and Culture* 2 (May-June): 16–18, 35.
Plato (1914). *Phaedo*. H. N. Fowler (trans.) in *Plato I*, Loeb Classical Library vol. 36. Cambridge: Harvard University Press.
Percy, W. (1975). *The message in the bottle: how queer man is, how queer language is, and what one has to do with the other*. New York: Farrar, Straus and Giroux.
Percy, W. (1991). From fact to Fiction, in *Signposts in a strange land*, P. Samway (ed.). New York: Farrar, Straus and Giroux.
Percy, W. (1998). *The moviegoer*. New York: Vintage International.
Polkinghorne, J. (1986). *One world: the interaction of science and theology*. Princeton: Princeton University Press.
Post, J.F. (1991). *Metaphysics: A contemporary introduction*. New York: Paragon House.
Post, J.F. (1987). *The faces of existence: An essay in nonreductive metaphysics*. Ithaca: Cornell University Press.
Quinlan, K. (1996). *Walker Percy: the last Catholic novelist*. Baton Rouge: Louisiana State University Press.
Rudnicki, R.W. (1999). *Percyscapes: the fugue state in twentieth-century southern fiction*. Baton Rouge: Louisiana State University Press.
Sartre, J.-P. (1948). *Existentialism and humanism*, P. Mairet (trans.). London: Methuen.
Stewart, H.F. (1974). *Boethius: an essay*. New York: Lenox Hill (reprint, London: W. Blackwood, 1891).
Todd, O. (2000). *Albert Camus: A life*, B. Ivry (trans.). New York: Carroll and Graf (reprint, New York: Alfred A. Knopf, 1997).
Wood, R.C. (1988). *The comedy of redemption: Christian faith and comic vision in four American novelists*. notre Dame: University of Notre Dame Press.

Preparation for the Cure

Anthony E. Giampietro, CSB

1 Introduction

The question of the role of philosophy in the life of the person of faith has engaged the greatest minds for millennia. And the question of whether Anicius Manlius Severinus Boethius was a martyr for the Christian faith has also engaged great minds. In *De Consolatione Philosophiae*, Boethius mentions almost nothing about Christianity and about the place of Christian faith in the sufferings he was enduring. "Lady Philosophy" comes to console Boethius, comes to answer his questions, to provide help in his hour of need. That she does not free him physically from his imprisonment makes it clear that the consolation she provides is an *interior* consolation, a consolation that comes from within or that is experienced from within. Lady Philosophy argues that "God and true happiness are one and the same" (Boethius, 1962, p. 65), and this makes the issue of disease and cure a profoundly interesting one as it pertains to philosophy and theology. And it is important for persons of all time periods, for if it is philosophy alone that cures, then our salvation is found in rationalism and not in faith.

Professor Douglas Henry provides a very thoughtful reflection on the role of philosophy in the life of Boethius. Throughout his discussion of Boethius and philosophy one is provided the opportunity to see more clearly the specifics of Boethius's struggles with the disease(s) of bad thinking. Dr. Henry presents very well the similarities between what Boethius suffers and the intellectual challenges of the 21st century. In the process Henry hopes to show that "philosophy"—faithfully followed—"can both console and cure us" (2009, p. 78).

In this chapter, I offer some suggestions to improve Dr. Henry's thesis. Good philosophy, philosophy that is not merely self-affirming therapeutic drivel (p. 78), helps one to see one's actual *source* and *end*. And while it can "provide morally significant direction for our lives," it is unable to cure.

A.E. Giampietro, CSB (✉)
Department of Philosophy, University of Saint Thomas, Houston, TX, USA
e-mail: aegcsb@stthom.edu

M.J. Cherry (ed.), *The Normativity of the Natural*, Philosophical Studies in Contemporary Culture 16, DOI 10.1007/978-90-481-2301-8_7,
© Springer Science+Business Media B.V. 2009

2 Human Nature

Is there a human nature? If there is not, then each person builds his life entirely as he chooses. If there *is* a human nature, then each human being is bound in some way to something that is "given," something discovered, something received "from outside." Without the sense that one's nature is given and not created anew by each person, one can feel adrift in the universe, uncertain that there could be any meaning or purpose. Professor Henry notes well that the *telos* of man tells us what he *is*. To misunderstand my *telos*, my *end*, is to misunderstand what and who I am. In traditional *natural law* thinking, the human being has potential for a *telos* that is given from above. I leave aside here the question of whether the end itself is "in" human nature or "above" it.

Contemporary thought presents at least two serious challenges to this traditional understanding of human nature and purpose. The first is to dispute the claim that there *is* a stable human nature, a nature to which every human being is in some sense *bound*. There are a variety of reasons given for this view, from an emphasis on radical human freedom to a rejection that there are such things as "universals." There are political manifestations of this view. There is a rejection of the view that politicians have any grounds for making robust claims about human nature and about how humans ought to act. And there is a fear that any description of "Human Nature," far from being representative of a universal *truth*, amounts to a sectarian understanding that is imposed upon others.

A second challenge to traditional natural law thinking is the view that even if human nature is a *something* that is received-not-created by each person, it is nevertheless not a nature about which much can be known. By this line of thought, similar problems emerge as those noted above. Hence, to the statement of Lady Philosophy: "You have forgotten what you are," the contemporary response is either "There is no *what* that I am," or "What I am is *not knowable* by me or you."

Professor Henry presents Boethius as someone who holds the view that human nature is given and not created. Each human being is not radically free to make himself whatever he would like to be. However, not everyone is adequately attentive to his or her nature. So Professor Henry focuses his reflections on Lady Philosophy's statement to Boethius that he has forgotten who he is.

Professor Henry suggests to us that human beings can lose their natures (p. 88). This is surely a view that must be reckoned with, and not just by Boethius. Current biotechnology presents critical challenges to the view that human nature is *stable*, that it is something that cannot be changed. If we can modify human genes perhaps we can cause human beings to lose their natures. If we do this, it may be said that we have created new beings with new natures, or simply "different" beings, if one holds that *nature* is a theoretical construction. It goes without saying that if we have lost our natures then we are truly lost. If Boethius has lost his rational nature then philosophy would be unable to cure him.

Dr. Henry notes that Lady Philosophy intends to diagnose and treat Boethius in order to restore to him proper reason and, she hopes, thereby happiness as well (p. 88). Henry holds the view that we postmoderns are in a similar situation as Boethius except that we have willfully rejected truths that Boethius can still build upon. At least Boethius acknowledges "God as creator, the world as meaningfully ordered, and humankind as rational and mortal" (p. 78). In other words, we are, or should be, in deeper despair than Boethius. Focusing on what he calls the chief sources of discontent, *scientific rationalism* and *misologistic nihilism*, Henry hopes "that Lady Philosophy's therapeutic responses will provide for us a restorative remedy to which we can turn and to which we can direct others" (p. 78).

Boethius recognizes himself as a rational animal, and nothing more. Henry shows that the contemporary philosopher Daniel Dennett holds a similar view, and that he too has a desire for "something more," at least as it pertains to living morally. But Henry explains why the philosophy espoused by Dennett cannot *give* him something more; it cannot offer him more than a poetic hopefulness, "nonsense on stilts," to use Dennett's own words. Dennett's conviction is that the *real* is what is uncovered by the hard sciences, by investigation into the physical properties of things. For Dennett, Henry writes, "[e]verything else is nothing but an epiphenomenal ripple on the surface of a physical substrate" (p. 79).

Dennett recognizes that something more than physics is present, but it is not the kind of "more" that can be investigated. He notes that human minds are unique, but this uniqueness does not deliver a moral system. Boethius recognizes that living rationally does not guarantee happiness in this life, and he finds himself in some despair. Dennett recognizes that living rationally may at times require one to act against one's own interests. Whereas Boethius sees that this makes everything absurd, Dennett is content to plow forward without any deeper justification. As Henry writes, Dennett's account "does not go far enough; it does not embrace the 'something more' which Lady Philosophy regards as indispensable for a complete account of human morality" (p. 81). If moral behavior does not guarantee happiness, why be moral?

Professor Henry challenges Dennett's argument at its heart (regarding how we ought to live). For Dennett, moral oughtness just *is*. Rules are good because they *work*. But which "rules" are to be preferred? Upon what does obligation depend? On the whim of human feeling and choice, it turns out. Unlike Dennett, Boethius will accept that human beings are *something more* than just rational and mortal. He accepts that human nature "properly finds its meaning within and only within the framing context of its origin and end in God."

In my view, what Professor Henry does not adequately show is how philosophy has helped Boethius to accept this understanding, entailing as it does a rather robust grasp of metaphysics and of providence. Henry is quite right to highlight the fact that, at the end of the day, moral obligation is "nonsense on stilts" unless there is something more than Dennett's "rationality" and "genius." Such words give the impression that "someone" is in charge of it all; but this is precisely what Dennett denies. And "[w]here Dennett does not baptize the irrational, unjustifiable, but evolutionarily derived status quo as morally apropos, then, he instead holds out for an ideal account at a future point that sensibly offers meaning where none is now apparent" (p. 83). In short, Dennett tells us to do what seems right (for no defensible reasons) and to have confidence that some day we will figure it all out. But human beings are inclined to want more than this; the *Consolatio* offers confidence that we *can* have more.

3 More About "Something More"

Can philosophy console? Can philosophy cure? At a minimum, philosophy, done well, can help us to see things as they are. And to see things as they are is in some sense the cure for what ails us. Can we say that we can know by looking at human nature what is our end and purpose? Dr. Henry suggests that philosophy can both console and cure. I think that his analysis of Boethius's text gives us some reason to think more deeply about many of the elements of our own culture, specifically in relation to scientific rationalism and misologistic nihilism. And I think that his attempt to make us think more deeply about these issues in fact leads us toward the cures. What I suggest is missing is precisely how the "more" can be known philosophically. That is, in my view, more can be said about the *purpose* that philosophy can tell us about.

Coming to realize that God is our source and our end implies that there is "something more that can be known." Daniel Dennett's scientific rationalism takes the place of Thomistic metaphysics, a metaphysics that is a requirement both for natural law thinking and for answering questions to which Dennett himself adverts. Once one abandons hope of discovering the nature of reality, unity and diversity, the nature of the source of existence, one is left adrift in the cosmos. As Henry notes, Dennett privileges a certain type of metaphysics, a materialism that is incapable of "something more." Scientific naturalism is taken to be "the way things are." Hence, while Dennett wants there to be human rights, he cannot find grounds for these rights in reality as he has understood it. And if one examines the "hot topics" of public discourse, e.g., stem cell research, constitutional law, same-sex marriage, evolution, etc., one will understand how important is a retrieval of a philosophical approach adequate to the discourse. Professor Henry's discussion of Boethius provides a foundation for serious critiques of many widely held views. It is a critique of certain attempts to justify ethical or political claims, attempts that are inadequate because they rest on the conviction that metaphysics cannot deliver moral obligation. In the wake of such inadequacy, one is left with arbitrary preference and political power to enforce this preference or to fail to enforce it. That there is nevertheless a type of *faith* in scientific rationalism is indicative of the intellectual malaise of our time.

In his discussion of *Misologistic Nihilism*, professor Henry makes some very thought-provoking connections between Boethius's disease and contemporary nihilism. Despairing of what the intellect can accomplish, many people "abandon hope in reason and settle for skepticism, cynicism, or some kind of self-authorized meaning-making of their own." And this is not only in the halls of academia. In a well-known and oft-quoted decision, the United States Supreme Court ruled that "[a]t the heart of liberty is the right to define one's own concept of existence, of meaning, of the universe, and of the mystery of human life" (Planned Parenthood v Casey, 505 US 833, 1992). It is not my intent here to call this decision into question but rather to note that it indicates how thin is the public understanding of what it means to be human. Freedom to "be who I want to be," to "have it all," and "not to live by anyone else's rules" are not merely adolescent claims anymore; anything short of such freedom is thought to be oppressive.

So, in the end, can philosophy console and cure someone like Boethius? Properly understood, it can only set the stage for the cure. Good philosophy can provide reason for thinking there is unity and purpose in the universe and that there is a natural moral law that, if human beings follow, it will point toward a lasting happiness. And philosophy can thus give good reasons for acting morally, even if one must suffer death in doing so. But philosophy cannot cure. It can only prepare the person for the healing that must come from elsewhere.

Moral truths can be found only by examining more than human nature by itself. Human beings can know, not just *believe*, that God is our source and our end. By examining human nature we begin to see that other things "must be the case." What are some of these things? That a human being has an immaterial soul that comes from elsewhere than his or her human parents. That there is a primary mover, not made out of matter, that has no beginning, that is not made up of parts. That human beings have intellects and wills that are connected to the human body but whose activities also transcend the activities of the human body. That existence is good, that human beings cannot help but pursue their own happiness. And humans can discover things about the nature of friendship that tell us about the nature of the *Good*. Philosophy can also reveal the inadequacy of certain claims about morality, of some of the claims made by the scientific rationalist and the misologistic nihilist. The inadequacy can be shown by a demonstration that their claims can be reduced to *unreasonable* statements about reality. Thus by way of philosophy, we are exhorted to look elsewhere for the answers to some of our questions.

A moral theory that presupposes materialism and rejects teleology cannot help but leave the great questions of life unanswered. Philosophy can help us, Henry writes, when it "discerns *what we are* within the context of our *source* and our *end*" (p. 78). Henry contends that "knowledge of what human beings are provides morally normative direction for our lives." This is no small task, but it can be done. A comprehensive look at the proofs for God's existence and at the claims of those contemporary thinkers who are atheistic is essential to it. For example, in "The Second Tablet Project," J. Budziszewski examines the claims of those who deny the existence of God and argues that they forfeit their moral principles (2002, pp. 23–31). Budziszewski notes that Thomas Nagel rejects theism not because it is unreasonable, but because he "does not want the universe to be like that." Nagel knows that if the universe is "like that," then he would be morally bound to change how he lives. He exhibits what professor Henry calls "a willful giving over of, a desisting from, knowledge of what we are—and more" (p. 77).

For Henry, the something more "involves discerning that human nature, rational and mortal as it is, properly finds its meaning within and only within the framing context of its origin and end in God." However, this is not a truth that results from philosophy alone. In the end, it is not philosophy itself that cures. Even Lady Philosophy points beyond reason at the end of the *Consolation*. As Henry notes, using the words of Pope John Paul II, "the neglect of being inevitably leads to losing touch with objective truth and therefore with the very ground of human dignity" (Fides et Ratio, p. 90). The "intuition of being" is perhaps more like a *gift* than something that can be argued for philosophically. And it prepares us both for adequate moral reasoning and for salvation.

Acknowledgments Many thanks to Dr. Siobhan Marshall for her helpful comments as I prepared this chapter.

Bibliography

Boethius, Anicius Manlius Severinus (1962). *The consolation of philosophy*, R. Green (trans.). New York: Macmillan.

Budziszewski, J. (2002). The second tablet project. *First Things*, June/July, 23–31.

Campenhausen, Hans von (1964). *The fathers of the Latin church*, M. Hoffman (trans.). New York: Harper & Row.

Chadwick, Henry (1981). *Boethius: The consolations of music, logic, theology, and philosophy*. Oxford: Oxford University Press.

de Labriolle, Pierre (1924). *History and literature of Christianity from Tertullian to Boethius*, H. Wilson (trans.). New York: A.A. Knopf.

Henry, D. (2009). Quid ipse sis nosse desisti, in M.J. Cherry (ed.), *The normativity of the natural*. Dordrecht: Springer.

John Paul II. (1998). *Fides et ratio*. Vatican City: Vatican.

Marenbon, John (2003). *Boethius*. Oxford: Oxford University Press.

McInerny, Ralph (1990). *Boethius and Aquinas*. Washington, DC: Catholic University of America Press.

McMahon, Robert (2006). *Understanding the medieval meditative ascent: Augustine, Anselm, Boethius, and Dante*. Washington, DC: Catholic University of America Press.

Rand, Edward Kennard (1957). *Founders of the middle ages*. New York: Dover Publications.

Diagnosing Cultural Progress and Decline

William J. Zanardi

1 Introduction: Changing the Focus

Some critics of postmodern philosophy claim that the waning of classical culture and its metaphysical tradition has resulted in the "death of culture." Has there been such a death? Has the rejection of traditional metaphysics meant the deculturation of the West? Rather than focus on questions of morbidity, I think it more productive to focus on questions of genesis. For example, what were the origins of classical culture and what comes after it?

Borrowing from the work of Eric Voegelin, I assume there is a primordial experience out of which arise all cultures and from which, much later, may arise explicit metaphysical theories (Voegelin, 1974, pp. 74–75). This primordial experience has a three-part structure. Briefly sketched: persons experience things, including themselves, as coming into being and passing away; i.e., they experience the *transiency* of existence. But human beings are distinct in their capacity to wonder and to ask questions about themselves and their world. So the second term in the pattern is the *questioning* evoked by the experience of transiency: What sense is there to all this coming into being and passing away? The third term is the *response* sought by the questioning and eventually shaping a specific culture. A correlation of these three terms ($T \leftrightarrow Q \leftrightarrow R$) provides an analytical pattern for investigating any culture and its implicit or explicit metaphysics (Voegelin, 1974, pp. 74–75). The structure is a constant; the three terms are the key variables, much as distance and time are variables in Galileo's fixing of the meaning of velocity.[1]

How does this heuristic pattern relate to the previous questions about cultural death or deculturation? If a cultural worldview or an explicit metaphysics loses its plausibility for large numbers of persons within a society, its replacements are likely already present as competitors for that public's assent and loyalty. Why? Just because a specific "R" loses credibility does not mean the primordial structure disappears. The experience of transiency and the questioning continue; i.e., persons

W.J. Zanardi (✉)
Department of Philosophy, St. Edward's University, Austin, TX, USA
e-mail: williamz@stedwards.edu

M.J. Cherry (ed.), *The Normativity of the Natural*, Philosophical Studies in Contemporary Culture 16, DOI 10.1007/978-90-481-2301-8_8,
© Springer Science+Business Media B.V. 2009

continue to wonder about their own existence in time. It was the mistake of classical culture in the West to assume that its responses to questions about what was true, good, and beautiful were ahistorical meanings that were the measure of all other cultural responses. While classical culture has largely fallen from favor, the structural constant at its founding remains. So the cultural responses to the primordial experience and any implicit or explicit metaphysical answers are variables. Historically formulated responses may "die," but the search for meaningful responses endures and evokes new responses.

But amid all the coming into being and passing away of cultural responses, do we find any grounds for estimating when cultural changes are advances or declines? Perhaps treating cultural responses as historical variables leads us to a relativistic conclusion about the incommensurability of diverse cultural meanings and metaphysical worldviews.[2] In the following pages, I will propose an alternative conclusion, one indebted to the natural law tradition in the West and to the work of Bernard Lonergan.

The central puzzle is how to distinguish cultural advance from cultural decline. This is the *question of diagnosis*. That answering it presupposes an answer to a prior question may be apparent if we accept that diagnosing diseases presupposes at least an implicit understanding of health. In other words, any diagnosis of cultural advance or decline presupposes that one has answered a prior question about historical process: *Ideally, what goes forward in history?*[3] Only if this question is answered and the answer is taken to be normative, are diagnoses likely to be defensible. So the steps to take are threefold: (1) develop a normative account of historical process; (2) indicate how it may be used to assess cultural responses as less advanced or as more advanced; and (3) suggest how one might diagnose cultural decline.

2 Approximating a Normative Account of Historical Process

In keeping with the reference to Galileo above, a theory of historical process must aim at identifying key variables and the invariant patterns of relations among them. While descriptive narratives proceeding from the "bottom up" may collect numerous historical data, the selection and explanation of the data require an intelligent grasp of relevant patterns. Since the question is one of historical process, the relevant patterns cannot be static. So what key terms or variables can suggest a dynamic, heuristic account of historical order? A search for them begins with trial and error, and the proof of their validity comes later in the efficiency with which they order and illuminate vast amounts of historical data (Lonergan, 1999, p. 19).

Where to begin the search? As long as the results of the search are open to revision, the starting point need not be justified as more than an intelligent first approximation to an answer.[4] Suppose we begin with a traditional theological pattern and assume that historical process results from interaction among three distinct realities: *nature, culture and God*.[5] The natural law tradition makes the ultimate agent of historical process a divine orderer. The absolute measure of all world processes, and so of human progress, is a divine Creator. Still, nature has its own ends. Biological processes recurrently pursue the emergence and survival of a species. In their routine functioning, these processes supply the conditions for the development of even higher-level activities. Thus, hunger serves as a stimulus to food gathering (i.e., a routine means to a natural end), but, in time, eating may become a ritual of family dining and gourmet cooking. What was once biological necessity may become cultural activity taking a variety of forms. Human history as culture-building begins to be distinct from nature when persons turn biological necessities into the varied forms of making dinner, shelter, clothing and war.

So there is a horizontal process of biological activity through which a species preserves and replicates itself. Think of the routines of a rabbit searching for food, a mate and security. But there is also a vertical process whereby routine actions at a lower level provide the materials from which higher-order activities emerge and by which they are sustained. For example, biological development in the adolescent gives rise to opportunities for dating rituals, dance clubs, wedding planners and laws stipulating eligibility for marriage. Institutions are not biological entities, so a dynamic view of history expands beyond the term "nature." A quick reference to artistic activity may solidify this basic insight. The raw materials for the sculptor, whether they be stone or metal, are sublated into a vertical ordering that could not be possible without them but which is not reducible to them. In addition, there is the testimony in stone and song and image that cultural artifacts respond to a divine impetus. The towers of a cathedral, the music that fills its spaces, the images in its windows express a further type of verticality. Push this vertical ordering far enough and cultural-historical process may be thought of as an unfinished drama of divine-human interaction.

This quick sketch of three interrelated terms provides a first approximation to a dynamic version of cultural advance and decline. The three terms provide a heuristic framework of sufficient generality to begin ordering any number of historical data irrespective of the time or place.[6] At a minimum some such framework will be part of any historical research since data neither speak for themselves nor do they assemble themselves into meaningful orders.[7] Whatever scheme is adopted for an initial framing of the data, the user must show the device does not undermine the openness of the inquiry. Keeping the basic terms at a level of great generality is one way of protecting the openness of the inquiry. Again, the alternative to "framing devices" or models composed of sets of related terms is not inquiry without any such devices but inquiry that randomly collects fragmentary pieces of information.

The first model for ordering the data is dynamic in the sense that higher-level processes pursuing higher ends are based on the incidental conditions of lower-level processes which pursue their own proper ends. A useful analogy is how the complex processes of sight provide materials for intellectual inquiry. Seeing has its end in images, and images in turn can become the focus of questioning; but questioning has its proper end not in more images but in the event called insight when one comes to some understanding of what it is one is seeing.

To refine the first approximation to a dynamic model of historical process, let me suggest a second set of terms parallel to the first three terms of nature, culture and God. This second set of terms identifies three different ends of persons: *life*, the *good life* and *eternal life*.[8] Nature corresponds to life in that both exhibit repetitive cycles of emergence, maintenance and death.[9] In contrast, the good life exceeds biological needs and their temporary and repetitive satisfaction. While economic activities transform material conditions of nature into the means for satisfying basic human needs, the good life, as a goal of human ingenuity and aspiration, elicits far more diverse activities than those needed to maintain life. "Culture" is the term corresponding to the good life because human aspirations do not settle for the repetitive but demand development and, in doing so, make living an unfinished drama. Aristotle said, "We are busy so as to have leisure," by which he meant that taking care of basic necessities is preliminary to the pursuit of higher ends (Aristotle, 1177b, 6). A wide range of cultural activities is made possible by meeting effectively and recurrently the demands of nature or life, but those cultural activities are not reducible to such practical matters.

Perhaps the contrast between the end of life and the end of the good life is more apparent when the third term is introduced. Eternal life corresponds to the earlier term "God." Translated into the earlier correlation of transiency, questioning and response, "eternal life" reflects the role in cultural history of responses that identified the process of coming into being and passing away as a process of displacement *away from* activities serving transparent goods and *toward* activities serving opaque but ultimate goods. Life's ends of survival and continuance of the species remain transparent goods of familiar daily activities. Technology and economic activities initially serve social ends by sustaining those institutions which provide systematic and recurrent ways of satisfying basic needs. But "bread alone" is not an ultimate end of living, especially for those whose material conditions allow for leisure. Historically, their questioning or wondering about transiency has not settled for responses that suggest living is an end in itself. But then the good life with its demand for growth toward higher ends may itself become the focus of questioning. If material satisfaction is not equated with the end of the good life, what else might? The responses are historically varied. The usual triad is: (1) a life of aesthetic enjoyment and adventure, (2) a life of moral virtue, (3) the journey of spiritual transformation.[10] All three responses are distinct from transparent material or economic goods.

To review the preceding remarks before moving onto the next approximation: we began with the primordial and most general pattern of $T \leftrightarrow Q \leftrightarrow R$. The first approximation offered a more determinate version of this cultural constant by linking three terms as correlates comprising a response (R): nature, culture, God. The second approximation further specified these distinct but related terms by distinguishing human ends: life, the good life and eternal life. Already ingredient in both approximations was a notion of hierarchy and, hence, a normative understanding of human growth or progress. Far from being neutral descriptions of historical process, these patterns presuppose an ideal line of development. They are, in Aristotelian terms, supplying the "form" of history, the intelligible pattern of historical materials. Suppose that all understanding operates this way in ordering otherwise fragmentary data. Exploiting this understanding of how intelligence correlates data or unifies disparate clues suggests a third approximation to a model of historical process.[11]

Aquinas observed that, "The human species is a capacity to understand which proceeds to its perfection by way of incomplete acts of understanding" (quoted in Shute, p. 107). Suppose the "incomplete acts of understanding" can be generalized as *types of understanding* occurring in a developmental pattern. For example, highly variable common-sense understanding is what Socrates encountered among his Athenian audiences. In both their thinking and their doing, they were intent on the particular problems at hand and on how to remedy them. Socrates invited them to ask further questions that both exceeded the competence of common-sense understanding to answer and required a disengagement from the immediate concerns of practical living. They were asked to enter a realm of theory and to think about generalized relations among terms and actions. They were invited to reach for abstract patterns of relations which would explain why everyday experiences were what they were.

Such an invitation need not be accepted. For example, while knowing how to boil water is useful in daily living, knowing precisely why it boils is not required for survival. So common-sense living can give way to theoretical inquiry, but most civic leaders have better things to do with their time, and, besides, their common-sense thinking may not see any benefit in the effort. In the case of Socrates, the audience resented the invitation and condemned the inviter. Subsequent generations took up the challenge to enter into a world of theoretical reflection, but common-sense living remains resistant to theoretical inquiry. For example, public speech continues undisturbed in its talk about the sun rising and setting. Despite the discoveries in psychology of perception, common sense has no doubt that seeing what is out there in front of us is the hallmark of knowing and of what is real.[12] More troubling examples are common-sense estimates of foreign policy in terms of *Realpolitik* and of economics in terms of inevitably competing egoisms.

So we have two stages of incomplete historical development: (1) common-sense living and thinking, and (2) common-sense living and a world of theoretical reflection alien to but alongside of common-sense thinking. Can we anticipate a third stage in which theoretical reflection would provide a guide to and intelligent control over common-sense living? In other words, human practice would be intelligently directed by theoretical reflection. If this seems vague, consider how the history of politics or economics has been a distressing record of pre-theoretical activities randomly responding to challenges, sometimes with intelligent decisions but oftentimes with transitory fears, ambitions and power games. Can one imagine a future in which theory and practice would be complementary, when the best available understanding was the widely accepted measure of what can and should be done? From this perspective, the political and economic realisms of recent centuries up to the present will seem adolescent rantings and inept adaptations of the human species to its problems.

From this approximative standpoint, *cultural progress* is the slow acceptance of the demands of intelligence for more complete acts of understanding, demands resisted by the inertia of unintelligible situations and the habitual solutions of prevailing common sense. Cultural decline occurs because many reject these demands and instead accept some inherited nonsense as the best they can do. Notice this diagnosis rests on normative grounds, and arguably they are part of natural law. In particular, do we find in ourselves a basic imperative to strive to understand correctly and to act in ways consistent with our best available understanding? If so, then willingness to act intelligently and actually doing so are preconditions to progress because our capacity to understand naturally "proceeds to its perfection by way of incomplete acts of understanding." Put more concretely, the slow emergence of "better times" depends on detection of mistakes, a willingness to reverse their effects, discovery of more promising alternatives and a recurrent willingness to pursue them. Yet, these preconditions need to be fulfilled not only in a few persons but in many. As one generation passes on to the next what it has learned and achieved, so the flow of history accumulates both the nonsense and the good sense of what many have understood, willed and done. Progress by the solitary individual either shapes the cumulative understanding and decisions of the many or it is lost. Perhaps it will be rediscovered by a later generation, but in the short run it is likely to be an incidental deviation from the general run of cases.

The implication here is that the proper focus of the study of cultural advance and decline is not the individual but the species. What the creative few achieve may be instrumental in reshaping, for good or ill, the direction of the many, but the path opened up by creative individuals may be neglected by many; it may become a road not taken. Consider how political revolts attack a privileged class. Creative individuals may envision a society in which the evils inflicted by a once powerful group need not be repeated by their successors. Whether their vision is accompanied by concrete proposals for reversing the exploitation of one group by another is less important than whether the successors have detected and are willing to break the old cycle of changing riders but keeping the same horse. That is, are they aware of the pattern of factional strife wherein subordinate groups supplant dominant groups and then do unto others as has been done unto them? Do they understand how to break this old cycle? Are they willing to try?

Absent this understanding and willingness, political realism (*Realpolitik*) and the acceptance of power games remain the prevailing "wisdom." Then cultural history becomes a cycle not far removed from the bloody vendetta. What can break that cycle is not the appeal of political expediency or common sense to "put the past behind us." Memories of injustices and losses are not so easily discarded. They will linger, fester and re-emerge as situations deteriorate and the old animosities are rekindled by new misunderstandings and injustices. Here is the place in the human story to make a choice. By itself theory seems ineffective in moving persons to forego the cycles of violence. What is needed are the affect-laden symbols that stir persons to wonder about the mystery of their existence in time.[13] So one formulation of the options can be put simply: either historical process is a story of ongoing violence interspersed with periods of peace or it is a story that includes something like the earlier terms of God and affective and effective intimations of eternal life as offering hope in living. Choosing between these alternatives actually reveals what one expects from historical process and how one conceives of historical progress.

3 A Fourth Approximation

A fourth correlation of terms suggests the role that hope or trust plays in the intelligent control of historical process. Suppose there is a dialectic of history arising dynamically from interaction among three realities: *historical context, human intelligence* and *horizon*.[14]

"Historical context" refers to an understanding of how decisions and actions of previous generations set the initial conditions for those of the next generation. A current situation is a flow of events that initially is an inheritance presenting a new generation with concrete problems to which it responds and advantages which it can exploit (Shute, p. 87). "Human intelligence" represents the precondition to historical progress insofar as bequeathed problems and advantages are real possibilities, but what is made of them is dependent on new decisions and actions. The first term then stands to the second as challenge stands to response.[15] So progress occurs if human intelligence continually meets problems with intelligent solutions and exploits advantages by creative improvements. While mistakes and accidents may make progress discontinuous, reversal of mistakes and recovery from accidents are part of human potential. The drama of human history is for this reason more than the repetitive cycles of nature. Put another way – the intervention of intelligence is what gives humankind an active role in making history (Shute, pp. 87–88). So the first part of this fourth pattern is a correlation of inherited challenges and responses to them, ideally by intelligent agents of historical progress.

With just these two terms, historical process can be understood as analogous to an experiment. Responses to inherited conditions will proceed by trial and error as persons pursue improvements. Why by "trial and error"? Consider how "messy" historical process tends to be. Inherited contexts will contain the results of intelligent choices and the lingering evils of unintelligent and even sinful choices. There will be practical insights on how to keep things moving forward, but there will also be habits of thought that resist adaptation to new circumstances, often by defenders of institutional orders whose primary concern is preserving positions of power. If you add to this the indeterminacy of actual situations and the pressure to react before persons have understood the challenges they face, the complexities become more apparent.[16] Progress comes, then, neither easily nor automatically. Its measure lies not in intentions but in outcomes, i.e., in whether persons succeed or fail in meeting challenges (Shute, p. 53).

What does the term "horizon" add to the preceding account of historical process as a tension between inherited context and human intelligence? Suppose by "horizon" is meant the range of questions and concerns a person finds worthy of attention. They need not be questions that same person can answer or concerns that he or she can adequately promote; it is enough if the person thinks they are significant

(Lonergan, 2001, pp. 298–299). What persons bring to the study of current events and problems will initially depend on the range of questions and concerns falling within their horizon. If the horizon of inquiry is too narrow, then relevant questions will remain unasked and so needed insights will not occur. Similarly, if the range of concerns is too narrow, legitimate goods (e.g., the well being of marginalized groups or environmental stability) may be neglected and even denied. If large numbers of persons operate within similarly narrow horizons, the historical context they pass on to the next generation will contain inadequately diagnosed problems, ineffective remedies and possibly worsening crises.

It would seem that part of the remedy to cultural decline must involve some kind of expansion of horizon.[17] Given the understanding of horizon as a range of questions and concerns thought significant, one would expect the horizons of different persons to be quite variable. Suppose the relevant variations fall under three distinct classes of questioning about historical process. The first type of questioning occurs when persons "see" no further than the concerns of practical living. This is the horizon of common sense in which questioning and concern ever revert back to how something will affect the persons doing the inquiring. The second type of questioning embraces a wider perspective in that questioning occurs about one's times and tribe. Here the questions need not always have a practical purpose though they often will focus on the defense of some group's interests. Still, they may be questions about how to promote progress within that group and may proceed from an understanding of the role of theory in guiding common-sense living. Finally, the third class exhibits theoretical questioning about the historical process in which entire civilizations come into being and pass away. The responses found within the first two horizons will be inadequate answers to the questions arising in the third since here the questioning and caring about meaning are not limited to concern for the good of an individual or of a particular society. The puzzling is about what sense historical process may make and what historical progress may mean.

From this third horizon more limited cultural responses will seem deficient. For example, Classical Liberalism in the West described society as a field of competition among self-interested individuals. With faith in the "hidden hand" of free markets, its advocates often saw no reason to worry whether individuals and institutions needed to grow morally in order to improve themselves and historical process.[18] Similarly, Marxist critics of the liberal state trusted in a hidden mechanism to remedy the ills of society. Contradictions between economic practices and human aspirations would eventually put an end to artificial obstacles to human progress. But such appeals to impersonal necessities are at odds with the basic notion of historical process as a dynamic experiment pursuing indeterminate ends. Neither the liberal state nor its rivals trusted that human intelligence could meet the challenge of open experimentation under conditions of uncertainty.[19] Instead, they placed their trust in something else as the agent of progress.

To emphasize the role of human intelligence in making history may seem to be an endorsement of the Enlightenment dream of rational persons building the heavenly city on earth. At a minimum we need to accept that the meanings people live by, their ideas and ideals, shape their choices and worlds. No one disputes this when the issue is technological innovations. Why should it be any less true when the issue is the breadth of their historical perspective?

To return to the question of hope and trust – what role does either play in the intelligent control of historical process? When our inherited context contains the accumulated absurdities of the past century and the early years of the new millennium already manifest new forms of violence, what response can intelligent persons make? Have we made much progress or is nonsense in the lead? If the horizon we rely on envisions only the last century, the lead belongs to nonsense. The experiment has not been going well, and western civilization takes on the features of a cultural slum.[20] But an expansion of horizon may counter this gloom.

What breadth of horizon is needed? Suppose the human journey has been a slow climb with occasional ascents to higher ground but also with periodic avalanches that sweep away prior achievements and leave survivors below where they began their ascent. Still, what is constant is the interplay of puzzling experience and inquiry into what it all may mean. The field intended by this questioning is in principle all that there is to be known and to be made.[21] What is actually known or made is far short of what the inquiry intends, yet the historical process is both the paltry achievements and the striving to surpass them. What inspires such striving? What is the basis for hope and trust that the climb is worth the effort?

4 A Fifth Approximation

To take the question of hope and trust seriously, let us make it part of a fifth approximation in this search for a normative account of historical process. Suppose we think of history as a dynamic process involving three terms: *creative intelligence*, *resistant biases* and *redemptive acts*.[22] The second term needs little explanation if by bias we mean the flight from further questions about what is the case and what should be done. Hegel summed up its recurrent appearance and effects by his description of history as "the slaughter-bench at which the happiness of peoples, the wisdom of States, and the virtue of individuals have been victimized...." (1956, p. 21). The last century with its wars and ideological struggles did nothing to change that early nineteenth-century appraisal. But the first term has its role in the process, so the last century saw innovations in dance and physics, in sculpture and genetics.[23] What is the role of the third term in relation to the other two?

As noted above, an actual historical situation results from decisions people have made. Those decisions may have followed upon careful efforts to be intelligent and responsible, but they may also have sprung from failures to be intelligent and responsible. Thus, we inherit complex situations today that include the consequences of bias, stupidity and moral fault. No surprise perhaps since, besides development, there is also decline. But, if the effects of decline are apparent in the nonsense of what the political realists may call "facts," there still is the possibility of development.

Suppose that development is of two kinds. Imagine two vertical arrows, one pointing up, the other down. The first image suggests how development can be from "below upwards." We sometimes creatively puzzle about our experiences, arrive at understanding, form sound judgments, and make responsible decisions. Doing so changes us for the better. The second arrow represents how we experience changes in our lives that originated from "above downwards." The easiest examples occur when others give us the gift of their prior learning. They hand down to us what they know but which we affirm as a matter of belief. How much of what anyone claims to know is in fact taken on faith? But this type of development also occurs as a result of love.

As part of human development, most persons experience a spontaneous caring about others that expands in time to include wider populations. The initial shift is from a concern for personal satisfaction to a concern for providing others with a better time, where at first the "others" are usually family members. Children learn to take responsibility for decisions affecting them. Eventually love of community and respect for persons as persons may be directing how persons respond to their own desires and how they decide to live. Love of others in these cases becomes a guiding form of caring that can lead persons to root out their own biases, to dismiss feelings of resentment and to resist calls for retaliation. As a result, they may avoid adding to and may even diminish some of the accumulated nonsense in their society; their development may play a part in reversing the effects of social decline.

A central theme of many religious traditions has been the role of religious conversion in changing lives for the better. Is it an instance of development from above downwards? In St. Paul's wording, religious conversion is the experience of God's love flooding one's heart (*Romans* 5:5). How might this "gift" make a basic difference in a person's thinking and acting? As makers of meaning, persons create the worlds in which they live; but, as sources of nonsense, they also infect those worlds with cycles of violence and malice. Insights can accumulate and produce progress; the effects of unintelligent and irresponsible choices can accumulate and produce bitter memories of betrayal and injustice. So in living with one another, persons need intelligently creative ways of "healing" the effects of past and current evils. But, in contrast to indifference and despair, the healing efforts require both caring and hoping. Does recognizing and returning God's love provide a "higher" perspective on such challenges, one that can transform anger and despair into forgiveness and hope? Can religious conversion liberate creative intelligence from the discouraging "facts" accepted by the political realist and orientate persons toward providing a better time for others?

These brief remarks on religious conversion suggest a role for the third term, "redemptive acts." In actual historical situations suppose there are three basic elements: *creative intelligence* intent on understanding the situation and finding ways to improve it; *biased* or *resistant intelligence* intent on defending some nonsense against the needed insights and decisions; *intelligence hopeful* that persons can live up to their own inner demands for being intelligent and responsible. The creative element is the spontaneous caring about understanding and about others. The biased element is the misdirection and underdevelopment of that caring, so that evils endure and accumulate. The healing element is the first term, creative intelligence, given hope in the face of those evils.

What examples do we have of the last term? Among the historical figures who come to mind, some will be explicitly religious persons. Martin Luther King, Jr. and Gandhi will occur to many as examples of hope-filled healers in the last century. The figure of Archbishop Desmond Tutu and his "hopelessly unrealistic" proposal of a Truth and Reconciliation Commission will occur to others. What evils did they find creative ways of resisting? Why were they hopeful despite the odds? And the odds are not good:

> We do not know ourselves very well; we cannot chart the future; we cannot control our environment completely or the influences that work on us; we cannot explore our unconscious and preconscious mechanisms. Our course is in the night; our control is only rough and approximate: we have to believe and trust; to risk and dare (Lonergan, 1967, p. 242).

Trusting and risking are acts that occur only when there is some hope of success. Even addicted gamblers have their fantasies. So what reasons do any of us have for believing and daring, for living intelligently with the mystery of human history?

5 Promises Kept?

Three promises were made in the Introduction: (1) to develop a normative account of historical process; (2) to indicate how it could be used to assess cultural responses as more advanced or less; (3) to suggest, as a result, how one might diagnose cultural decline. The first promise was kept even if doing so seemed to follow a twisting path. In working out a normative account through a series of five "approximations," I was imitating the strategy of Bernard Lonergan, a thinker absorbed by the question of historical process.[24] The chart in the Appendix offers a review of these five correlations of general categories, shifting from the most general terms (e.g., "response," "culture," and "the good life") to a less general classification of types of understanding (common sense and theory), and finally to highly variable yet ultimately concrete terms, e.g., "horizon," "historical context" and "redemptive acts."

The chart depicts a "shift to the right" as an indicator of how to assess cultural responses. So, for example, the shift of "the good life" toward the emergence of theory amid common-sense living is taken to be an instance of cultural advance. Few would doubt this judgment given the practical benefits of lower infant mortality rates from childhood diseases and the proliferation of scientific discoveries over the past three hundred years. The shift becomes more controversial when the term "horizon" becomes the key category in the fourth approximation of historical process. Three insights are the basis for this judgment of cultural advance: (1) human intelligence advances by raising and answering more questions; (2) the role of theory in guiding common-sense living marks an advance because the questions raised in theorizing are more comprehensive than those occurring to common sense;[25] (3) the corrective to biased intelligence lies not within common sense (which inevitably is some mixture of good sense and nonsense) but in some higher or more comprehensive understanding of human living.[26]

Cultural advance, then, proceeds by satisfying several conditions: (1) the slow emergence of more comprehensive questions about historical process; (2) the development of theoretical responses required by those expanded questions; (3) the randomly occurring (and potentially discontinuous and so non-cumulative) insights and actions of creative individuals; (4) the survival of those insights as eventually the possession of more than the few; and (5) increasing acceptance of the conviction that the inherited ills of any given historical situation require responses from creative intelligence informed by trust and hope.

Cultural responses that ignore one or more of these conditions are less advanced than responses that do not. For example, accounts that locate the *telos* of historical process in the intra-worldly success of a particular civilization, nation or economic system either ignore or beg the question about why this group or institutional order and not another should be the highest achievement of human creativity. Why should so much coming into being and passing away come to its end at this level? Usually the question is ignored as beyond the capacity of common sense to answer since it requires a shift in mentality from the palpable and imaginable objects and events of everyday living to the constructed patterns of relationships among events, that is, to the abstract correlations that comprise the content of theoretical understanding. Cultural responses that resist this shift fail to meet the first two conditions.

As further examples of less advanced responses, consider how relationships among persons and societies are fraught with tensions and inclinations to violence.[27] As a result, the whole intellectual enterprise of reaching for understanding may seem a very tenuous and constantly endangered activity. Only if intelligent inquiry is informed by hope will the efforts to understand what is and to create something better seem to be a reasonable gamble. But cultural responses that cite the pursuit of self-interest or the profit motive or factional strife as the basic driving forces of history will not include the need for *both* creative intelligence and redemptive acts in overcoming the accumulated nonsense found in inherited contexts. As a result, they are less advanced than cultural responses which give a central role to creative intelligence hopeful of overcoming the residual biases of the past and the animosities they spawn in the present.

But what of the diagnosis of *cultural decline*? Suppose it means a retreat from an earlier level of understanding and from patterns of concern appropriate to that level. As an example, think of the final page of Herodotus' *History of the Persian Wars*. Approximately twenty-three centuries before Napoleon, Herodotus portrays Cyrus as understanding that empire-building is a waste of time. Is there any need to comment on Hitler or Stalin as out of date well before their own times? Such negative examples underscore the first and third conditions of cultural advance, i.e., the slow emergence of questions and the tenuousness of the answers given by creative individuals.

As a further exercise in detecting cultural decline, note the moral resignation implicit in political realism, whether on the international level of *Realpolitik* or on the domestic level of factional power games. In both cases, advocates excuse immoral choices by appealing to the inevitability of such behavior, if not by them, then by their opponents. "The world is what it is, and not to adapt to 'ugly facts' simply invites one's own doom." But is this an apt example of cultural decline? If political realism has not been the exception but the rule in political history, how can it be a symptom of decline? Presumably "decline" presupposes a prior and higher level of cultural practice. One response is to recall the function of a normative account of historical process cited in endnote three. Insofar as the fifth approximation offers a general outline of such a normative account, the stance of political realism is deficient in its neglect of the role of creative intelligence and redemptive acts in countering the "ugly facts" of an inherited context.

Some closing words suggest an analogy between the preceding development of a view of historical process and Newton's development of an account of planetary motion.[28] Creative intelligence left undisturbed proceeds in a straight line to raise and answer questions and to resolve problems. The presence of resistant biases in individuals and in inherited contexts intervenes and deflects creative intelligence from the ideal path of development. When those interventions are recurrent and widespread, the "perturbations" convince many that historical process is absurd and creative intelligence is ineffective. In the face of these disturbances, many conclude either that history is a record of power games or "it is a tale told by an idiot, full of sound and fury, signifying nothing" and best responded to by quiet resignation. The first response invites one to be an accomplice in the nonsense, the second to abandon the making of history to the unscrupulous. Is there a third option? We end then by returning to the question of hope and trust.

Notes

1. Consider some possible variations in the content of these three terms. The experience of transiency varies among cultures since some will be relatively stable and others under constant threat of internal disorders or external invasion. The form questioning then takes may be as specific as why there are disorders in the empire or as general as Leibniz's query about why anything exists in the first place. Finally, as any study of creation myths makes evident, variations in the responses are abundant.
2. The literature on this topic contains at least two uses of "incommensurable." Sometimes it refers to an understanding that differences are irreducible but still compatible, e.g., choices between writing poetry or composing music. At other times what it seems to mean is that differences are both irreducible and incompatible, e.g., choices between policies promoting substantive equality among persons and policies preserving inequitable distributions of advantages and disadvantages. The second, stronger meaning of "incommensurable" is the focus of controversy since the "ends" pursued are both different and mutually exclusive.
3. An ideal account of historical process functions as a normative measure of actual situations. On the one hand, there is a rejection of a type of realism that takes actual situations as somehow normative in regard to human practice (Cf. the political realism associated with Machiavelli); on the other, there is an imagining of "history as better than it was" (Shute, 2010).
4. "Newton's planetary theory had a first approximation in the first law of motion; bodies move in a straight line with constant velocity unless some force intervenes. There was a second approximation when the addition of the law of gravity between the sun and the planet yielded an elliptical orbit for the planet. A third approximation was reached when the influence of the gravity of the planets on one another is taken into account to reveal the perturbed ellipses in which the planets actually move.

The point to this model is, of course, that in the intellectual construction of reality it is not any of the earlier stages of the construction but only the final product that actually exists" (Lonergan, 1974, pp. 271–272).

5. Shute points out a parallel approximation in Lonergan's work (see, for example, p. 126). Comte is well known for his law of three stages: the theological, the metaphysical and the positivist. His pattern assuredly departs from any natural law tradition that assumes a theological end as the ultimate measure of historical process. In the following pages I will be sketching five approximative accounts of historical process, and, while the terms of the first two parallel Comte's categories of the theological and metaphysical stages, their meanings will not be discarded as symptoms of human immaturity. Thus, I will be disassociating my account from Comte's linking of the first two stages with the childhood and adolescence of the human species as it develops in time.
6. "It remains that [historians] can follow a middle course, neither projecting into the past the categories of the present nor pretending that historical inquiry is conducted without a use of human intelligence. The middle course consists in constructing an a priori scheme that is capable of synthesizing any possible set of historical data irrespective of their place and time, just as the science of mathematics constructs a generic scheme capable of synthesizing any possible set of quantitative phenomena" (Lonergan, 2000, p. 156).
7. "Dynamically a science is the interplay of two factors: there are data revealed by experience, observation, experiment, measurement; and on the other hand, there is the constructive activity of mind. By themselves the data are objective, but they are also disparate, without significance, without correlation, without coherence. Of itself, the mind is coherence; spontaneously it constructs correlations and attributes significance; but it must have materials to construct and correlate; and if its work is not to be fanciful, its materials must be the data. Thus thought and experience are two complementary functions; thought constructs what experience reveals; and science is an exact equilibrium of the two" (Lonergan, 1998, p. 5).
8. See Shute (pp. 128–129), for the presence of this schema in Lonergan's work.
9. The link between "nature" and "mere life" and the distinctiveness of culture from both are part of Hegel's famous Master-Slave dialectic in *The Phenomenology of Spirit.*
10. A notable example of the triad is found in Kierkegaard's *Stages on Life's Way*: the aesthetic, the ethical and the religious options for living.
11. For this third approximation I am indebted to the analysis of Lonergan's thought on historical process provided in Shute (pp. 106–107).
12. One can read the history of modern philosophy as a struggle to relate these two distinct ways of understanding and the two realisms they imply (see, Lonergan, 1992, pp. 11–12).
13. "Lonergan grasped that a difficulty in Plato was his substitution of a concept for symbolism. Human beings do not live by concepts. For Lonergan it is specifically the conjunction of Greek philosophy and Christian faith that opens up the possibility of the infusion of affective and symbolic components into a purely intelligent or philosophic notion of world order" (Shute, p. 79).
14. This fourth pattern of terms is adapted from Lonergan's lectures on existentialism published in *Phenomenology and Logic* (2001, pp. 302f.).
15. I have in mind here Toynbee's pattern of civilizational process.
16. A standard maxim in foreign policy captures this sense of the indeterminacy of actual situations. When crises are most amenable to effective intervention, the needed information tends to be unavailable; when the information is available, the crises tend to have worsened beyond the point of easy resolution.
17. One instance was suggested above when noting the resistance of common-sense thinking to the expanded horizon of theory.
18. These remarks are not intended as a criticism of Adam Smith but of epigones who selectively read his work. For a corrective to the usual reading of Smith on the benefits of free markets (see Muller, 1993).
19. Note the ambiguity of Kant's position on historical progress. The relevant texts are found in a collection of Kant's political essays edited by Lewis White Beck (1963). On the one hand, Kant seemed to affirm the need for morally good persons as a precondition to progress (pp. 18 and 21),

but he also affirmed that Nature conspired to produce a good society even out of a "race of devils" so long as they were intelligent (p. 112).
20. Lonergan remarked on the deterioration of an entire society to the level of a "cultural slum" (1972, p. 99).
21. The Scholastics' maxim about human capacity is worth recalling: *potens omnia fieri et facere.*
22. The terms are borrowed from Bernard Lonergan, "Healing and Creating in History" (1985, pp. 100–114).
23. Fifty years ago C.P. Snow popularized talk of "two cultures" and invited a reading of the relation between the sciences and the humanities as antithetical. As an alternative, can we think of them as products of creative intelligence marred by resistant biases and so both in need of healing?
24. His strategy is revealed indirectly by the remarks quoted in endnote four.
25. Recall the simple example of questioning why water boils versus knowing how to boil it. Answering the former is more comprehensive in that one abstracts from all the variations in how to answer the latter.
26. Earlier remarks on religious conversion were elliptical suggestions that the horizon of persons in love can be such a higher viewpoint transcending a "realism" resigned to a cycle of revenge, the "ugly facts" of factional strife, the bitter memories of injustices and a politics of envy and suspicion.
27. The works of René Girard are a source of insights into the genesis of social violence. See especially his *Violence and the Sacred* (1977).
28. See endnote four.

Bibliography

Aristotle (1970). *Nichomachean ethics*, W.D. Ross (trans.). Oxford: Oxford University Press.

Girard, R. (1977). *Violence and the sacred*. Baltimore: Johns Hopkins University Press.

Hegel, G.W.F. (1956). *The philosophy of history*. New York: Dover Publications.

Kant, I. (1963). *On history*. Lewis White Beck (ed.). New York: Bobbs-Merrill.

Lonergan, B. (1967). *Existenz and aggiornamento*, in *Collection*. New York: Herder and Herder.

Lonergan, B. (1972). *Method in theology*. New York: Herder and Herder.

Lonergan, B. (1974). *Insight revisited*, in *A second collection*. Philadelphia: The Westminster Press.

Lonergan, B. (1985). *Healing and creating in history*, in *A third collection*. New York: Paulist Press.

Lonergan, B. (1992). *Insight*, in *Collected works*, vol. 3. Toronto: University of Toronto Press.

Lonergan, B. (1998). *For a new political economy*, in *Collected works*, vol. 21. Toronto: University of Toronto Press.

Lonergan, B. (1999). *Macroeconomic dynamics*, in *Collected works*, vol. 15. Toronto: University of Toronto Press.

Lonergan, B. (2000). *Grace and freedom*, in *Collected works*, vol. 1. Toronto: University of Toronto Press.

Lonergan, B. (2001). *Phenomenology and logic*, in *Collected works*, vol. 18. Toronto: University of Toronto Press.

Muller, J.Z. (1993). *Adam Smith in his time and ours*. New York: The Free Press.

Shute, M. (2010). *Lonergan's discovery of scientific economics*. Toronto: University of Toronto Press.

Voegelin, E. (1974). *The ecumenic age. Order and history*, vol. 4. Baton Rouge: Louisiana State University Press.

Appendix

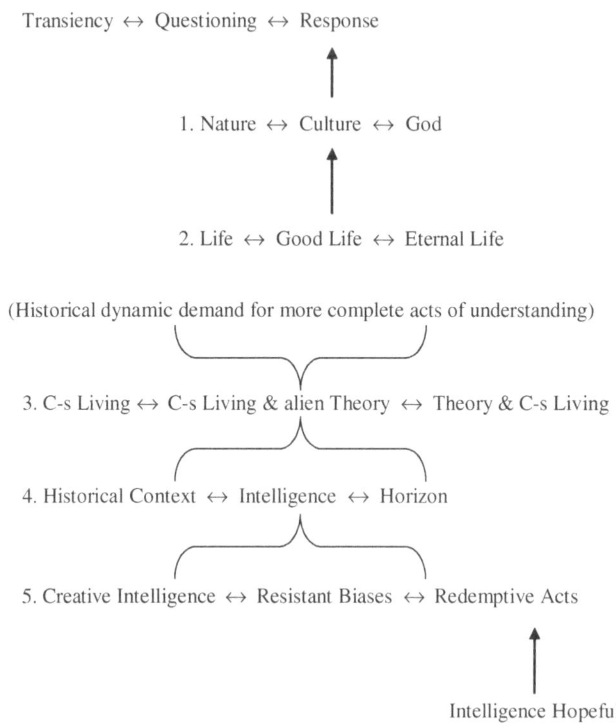
Transiency ↔ Questioning ↔ Response
1. Nature ↔ Culture ↔ God
2. Life ↔ Good Life ↔ Eternal Life
(Historical dynamic demand for more complete acts of understanding)
3. C-s Living ↔ C-s Living & alien Theory ↔ Theory & C-s Living
4. Historical Context ↔ Intelligence ↔ Horizon
5. Creative Intelligence ↔ Resistant Biases ↔ Redemptive Acts
Intelligence Hopeful

Part III
The Malleability of Human Nature

Reflections on Secular Foundationalism and Our Human Future

Stephen A. Erickson

Philosophy has come to some extraordinary and highly consequential cul-de-sacs in the course of its history. The most famous was during the period we label late Scholasticism. Undeterred by problematic assumptions and increasingly isolated methods, various scholastics refined and rigorously defended a number of exquisite theses. These thinkers were bent on besting each other in arguments remote not only from everyday concerns, but from the developing worldview of their intellectual peers in overlapping areas of investigation. These Schoolmen could rightly understand themselves to be defending important religious truths in the face of emerging threats from those we now routinely label as scientist. But many of these scientists were to their own minds equally religious in belief and commitment. This made it an odd time indeed. Its history was more complicated in the making than we can easily disentangle, though we still live in its wake.

Something similar, though neither as dramatic nor consequential, happened in early twentieth century England in the course of the transition from the Idealism of Bradley and McTaggert to the logical atomism pioneered by Russell and controversially transformed by the Wittgenstein of the *Tractatus*. A long dominant philosophical vocabulary withered in the face of a polemic that first attacked and then came to transcend and subsequently to ignore its vanquished predecessor. Internal relations and the Absolute, for example, gave way to external relations and a world of facts. The former notions became part of something called the history of philosophy or, more quaintly, the history of ideas.

A similar, though far, far weightier transition is occurring in our time as well. The players in the current drama, if we were to map them in terms of the journey out of medieval scholasticism, cannot but surprise us. We are offered somewhat of a reprise. It is as if the late medieval and early modern players have returned, retained updated versions of their faiths, but have also undergone a transfer of dominance, historical role, and to some extent even identifying jerseys. A group of very brilliant, secularly committed thinkers have been monopolizing and micro-

S.A. Erickson (✉)
Department of Philosophy, Pomona College, Claremont, CA, USA
e-mail: Stephen_Erickson@pomona.edu

M.J. Cherry (ed.), *The Normativity of the Natural*, Philosophical Studies in Contemporary Culture 16, DOI 10.1007/978-90-481-2301-8_9,
© Springer Science+Business Media B.V. 2009

managing issues ranging from human rights to cognitive perception, cultural values to questions regarding human dignity and meaning. This has been occurring in ways that have hermetically disregarded a great deal of the experience with which we are continually confronted, if not saturated. Yet, much of the currently established secular enterprise is so blinkered and counter-intuitive that it is able to continue only through remaining closed to all but those who are immersed in the underlying and largely shared assumptions of its professionally certified disputants.

In the meantime, as was the case during the late scholasticism and early modernism of the sixteenth and seventieth centuries, the surrounding world has been rapidly changing. Or perhaps in far less escapable ways than expected, this world is simply re-exhibiting some of its more ancient and surprisingly enduring features. The early twenty-first century is itself a very odd and perhaps also decisive historical moment. My hope is to capture a few aspects of its oddity…and promise…in some striking, albeit controversial ways.

There is little I can do in the face of our circumstances, except to forward some currently unfashionable observations and recommend some procedural consequence. I will highlight how little a secular conception of human nature has to offer with respect to the illumination of human conduct. My route to this end will be somewhat unconventional. This will constitute the major portion of my reflections.

To accomplish my ends I will be painting with some broad expressionistic strokes. Truly serious work in philosophy does require a significant measure of pointillism if it is to achieve traction and thereby gain sustained attention and professional respect. Of this I am well aware. But for such pointillism to become possible, a frame must first be provided and, thereby, a window offering a different way of "world-making." If pointillism does the eventual "making,"[1] it is nonetheless *framing* that renders that "making" first possible and credible. It is toward a new framing that my efforts are primarily directed.

Consider a thing-in-itself, a non-human and inanimate one that is what it is and not another thing. Let us skip past the claim that any such thing could only be known by means of a description and that description introduces insurmountable elements of conceptual mediation. In violation of the epistemological underpinnings of Kant's philosophy, for example, let us pretend to behold a thing-in-itself as it is in itself. What should follow from such a revelation?

An extreme answer would be that we could and then should confidently worship or at least obey such a thing. Less extremely, in our actions we could and should act in conformity with it. Surely this would be little more than to follow a primary dictate of scientific faith that is underwritten by faith in science itself: plan and act on the basis of how things certifiably are, not on the basis of how we might prefer them to be.

If the object of such a faith is *but* a thing, however, neither human nor even animate, it is hard to know how to follow through on this course of action. Would it mean for us to subordinate ourselves to the thing's interests, serving it through contributing to its preservation and conservation? This does have the ring of an environmentalist postulate. But could a thing of this sort plausibly be said to have interests? In order for these interests to be present—and thus contrary to our initial hypothesis—some sort of animation, some dimension of "life," might be required. Even animation might prove to be insufficient, however. That something is animated, even that it has interests, does not entail that these interests should be served. In fact, nothing follows at all with respect to subsequent human conduct, no matter how transparent and veridical our underlying and guiding perception might be.

Imagine an underlying continuum, on one end of which is strictly conforming obedience, even devotion to something. At the other end of this continuum, let us stipulate an unapologetic use of that same something to the point of its complete consumption, destruction, or extinction. It is quite possible to have diverse and varying preferences and/or temperamental dispositions with respect to the placement of items upon this continuum—in this limited sense "grounding" commitments that govern the locating and consequent relations we sustain to items coming to reside upon it. But, on the account so far provided, there is nothing guaranteed a priori, nothing in the "nature of things" to make these grounding dispositions or preferences right or wrong, appropriate or inappropriate—except perhaps in terms of their derivation from other, "higher" level dispositions or preferences.

To reach so-called "ultimate" dispositions or preferences, however, will simply be to arrive at a domain that is not itself subject to higher-level derivation. But from this formal circumstance nothing in particular will follow regarding the *value* of this "ultimate" domain beyond the inferential derivability of other, lower-level dispositions or preferences from it. That any domain could be shown to support and justify claims to inherent value—as opposed to having the mere significance of logically supporting other subordinate levels from which such problematic valuations arise—must, from the thoroughly secular standpoint I have been tracing, be construed as miraculous, i.e., as transcendent of secularity itself.

Let me put this more starkly. Thought through strictly, the worldview of any "unembellished," non-derivative secular humanism, thus of any humanism construed as a fundamental point of departure, cannot but have a bland and vacuous "nihilism" as one of its consequences. The only impediment to this conclusion would be a commitment or response that, however laudable, is unjustifiable and finally inexplicable on human terms alone.[2] At the same time, the sting is taken out of nihilism of this sort, for its articulation is merely an underwriting and labeling of a fact that it must embrace to be thoroughly consistent with itself, *viz.*, that what is...just simply...is, and from this nothing of positive (or negative) significance follows.

Those decrying views that lead to nihilism of this seemingly benign species are only able to do so through smuggling in a set of assumptions that the secular faith, the "scientific" humanism I have been adumbrating, must *ex hypothesei* find altogether foreign. Paradoxically, however, most secular and scientific humanists of the sort I have been describing abhor that bland nihilism that is not only consistent with but also required by their very position. They both do not want to acknowledge that any cake might be there to be had, thus their nihilism, but also want to munch on it anyway as witnessed by their discomfort with "nihilism." The consequences of the position they hold engenders a confusing hunger that is difficult for them fully to acknowledge. This is somehow very "Zen"—which in our current and commonly accepted vernacular is to say that it is at best opaque, at worst chronically bewildering, but that it at least thickens the plot.

Let me now turn more directly to a particular and far more relevant sort of "thing-in-itself," *viz.*, a person. This notion will shed light on the strangeness of that secular and scientific point of view I have been tracing. (Surely, some must think, I have caricatured or at least misleadingly described it.) Let us provisionally grant to a person, something that with one problematic exception is impossible to establish sufficiently in any interesting substantive or normative way: altogether secularly grounded traits that are universally human.[3] We now arrive at a disturbing question. What will follow with respect to human conduct?

One answer might be that we would know to what, in ourselves at least, to conform. But surely this is an unsatisfactory response. What would it mean, for example, to *conform* to universally human traits in ourselves? If the traits truly are universally human, we could not help but exhibit them in ourselves anyway. Conforming to such traits in ourselves would less be virtuous than matter of factly unavoidable.

Of course our plot really does thicken when we turn to that one problematic exception to my claim that universally human traits cannot be plausibly established by secular means. The one discoverable trait constituting the exception is that of possessing in principle, if not always in exercisable fact, personal freedom.[4] This is simply the directly experienced characteristic of having options and being free to choose among them. Any injunction directed toward us to conform (or not to conform) to any standard of any sort, particular or universal, presupposes this trait. But from the recurring practical experience of having the ability to make choices, little if anything follows with respect to what we actually can or ought to do.

For example, we might choose either to sustain other traits we discover ourselves already to have or resolve to transform or to eradicate some or even all of them. We might choose to diminish those traits that we experience as standing in the way of the full flexing of our freedom. But we might equally decide to strengthen some of these very freedom-constricting traits on the grounds that our freedom makes us unproductively anxious or is experienced by us in some stronger way as destructive to our sense of stability or well-being. Dostoevsky, among others, has written about such matters insightfully,[5] and it is clear that freedom for many can be as much a burden as an opportunity, more unsettling than liberating. That we discover this feature of ourselves tells us next to nothing regarding what to do with it.

A better approach to universally human traits, were they capable of being established, would involve the claim that they should be respected, at least in the minimal sense of not being violated. Knowing what these traits are would provide us with a checklist of items not to tamper with, at least not in others, and perhaps not in ourselves either.[6] But let us consider what I will somewhat playfully designate as wine, universality, and aggression—aggression first, universality next and then the grape.

It is contended by many that predatory, aggressive, and territorial instincts are innate features of humans, present in some more than others, but universally exhibited nonetheless. I need not rehearse human history nor recycle anecdotes from contemporary, garden-variety office politics to remind us that much of the "universally" human that we would refuse to violate on the recommendation before us would be tolerated at our peril. We often have to restrict the exercise of certain human traits simply to protect our interests and even our very selves. The underworld of Hobbes is never that far beneath our civilized and civilizing accomplishments and the order these accomplishments engender and precariously sustain. Survival itself sometimes requires dramatic and decisive interventions to prevent various all too human urges and activities from getting played out to the detriment of virtually everyone. That human nature is inherently benign, especially if appropriately nurtured, is a thesis chronically belied by our human history.

Consider next the notion of universality itself. It has been a problematic inheritance from the Greeks that we so easily construe universality as a mark of value and even conflate it at times with dignity itself. An epistemologically driven quest for the certain, universality's conversion into modernity made axiomatic through Descartes, combined with abiding reflections on those traits in us required to underwrite and secure certainty's achievement, have been entangled and fused with deep desires to establish a special status for ourselves *vis-à-vis* other creatures. Such a status has been meant to guarantee both our dignity and perpetuity.[7]

Immanuel Kant takes these conflations, reconstitutes them in modern form, and takes them to their seeming limit. It is against the backdrop of Kant's enlightenment rationalism, in fact, that the secular project has made most of its claims to progress. But let us consider the universal when contrasted with the particular and even unique, especially as we might employ these distinctions in the service of articulating various dimensions of persons. Wherein is the "valuable" found? Do we not invariably come down on the side of the unique and the individual, disconnecting these readily and decisively from epistemological desiderata regarding the comprehension of alleged universals? That we find someone unique, as opposed to being an instantiation of a universal—something we surely never do nor know how to achieve—overwhelms those compulsions toward universality that might in other domains dominate our considerations.

Let me attempt further to disentangle the conflations to which I have alluded and the central role that knowledge and, thus, through our Greek inheritance, reason has played in their formation and incremental fusion. Knowledge is taken to involve certainty and require a commonality of the knower and the known. What could be known with certainty, if not the permanent? But how could the permanent receive certification of its permanence except through its universality and thus transcendence of the realm of particularity, ravaged as the latter is by the erosions inflicted by chance and time? What then must follow with respect to the knower? Here the linkages are quite stretched, if no less decisive. If that which guarantees the certainty, and thereby the trustworthiness and security invested in the known, is its permanence, there must be something correspondingly permanent in the composition of the knower. As we know, this in fact is one of Plato's arguments for human immortality in the Phaedo.[8] On initial glance at least, we seem to be on potentially firm ground. The next linkage, however, is altogether problematic. Granting that permanence is secured through its attachment to universality on the "objective" side, that of the known, what meaning could be given to universality as a corresponding underwriter of the intelligibility and cogency of permanence on the "subjective" side, that of knowing and the knower? The answer that reflection offers, I think, is quite simply ... none, none at all. But if the linkages are truly necessary for requisite inferences to be validated, either human beings are in some fundamental way universal, not altogether personal and unique, or they fail the achievement of dignity through universality, thus failing to rise to that permanence—read as immortality—that the linkages are meant to secure.

The solution proposed, of course, was to locate the requisite universality in an essential feature of humans, *viz.*, the exercise of their reason. In this manner, the door is at least left open to assert the underlying uniqueness of the person. But the employment of this strategy is itself problematic. The quest for a principled, yet genuinely personal immortality was made insecure in its outcome. Salvation takes on an excessively cognitive coloration that flies in the face of most Western accounts of the spiritual predicament of existentially fallen human beings.

Let us look further into this crucial notion of reason, without which the Greek "problematic" would surely have foundered. Let us beat some closely neighboring bushes. So far I have concentrated primarily on the epistemological dimension of the "thing-in-itself" issue. Once, contrary to doctrine, a "thing-in-itself" is known, what if anything follows, I have asked, with respect to what ought to be done with or about it? Perhaps answers to this question become more readily available, if we consider the underlying "metaphysics" of the matter more closely.

We have long lived with an inheritance insisting that what distinguishes us from other creatures is also and precisely that which constitutes our "true" nature, our nature as it is in itself. We are further told that our distinguishing human feature—and thus our true nature—is intimately bound up in one particular activity, *viz.*, knowing.

A complexity is embedded in these Greek reflections that awaits the German tradition—most specifically Hegel—for its further illumination and disentanglement. On the one hand, knowing is that which enables us to *achieve* our true nature. But knowing, it is claimed, also *is* our true nature, and that about us that enables us to achieve this knowing, necessarily already in place, is itself claimed to be constitutive of that true nature as well.

What is it about us that enables us to know? Beyond the obvious reference to Reason itself there are a number of answers to this question. One is access to something *to* know, thus the capacity for receptivity.[9] Another is something in or about us that facilitates and supports the activity of knowing, as engine fuel, for example, facilitates and sustains the integrated operation of the complex circuitry of a dynamic engine. But we also inherit from the Greeks two, less obvious and far more controversial and intriguing assumptions. The first, as we have seen, is that knowing must share some features of the known, if it is to succeed and thus *be* knowing. For the Greeks, as we also know, this principle had a powerful underlying motivation. If the truly known was eternal and unchanging, and we could know it, then we must ourselves have an eternal and unchanging element that has allowed this knowing to occur—most plausibly an immortal soul. A second Greek assumption is that knowing increasingly assimilates the basic features of that which it comes to know.[10] Through knowing the knower becomes more like and in fact moves asymptotically towards identification with the known. At our core, thus, is a potentially extraordinary malleability. It contributes substantially to our capacity to know and to our subsequent "nature."

These two assumptions are built into and largely constitute the Platonic doctrine of participation. Its life has been much longer and more resilient than Platonism itself. Note that participation in this double sense helps explain why attacks on a particular community's culture—which is both a source and an object of especially intimate knowing—is unavoidably and altogether naturally experienced by that community as an attack on itself and its very identity. Given the underlying dynamics of participation, how could it be otherwise?

The linkages in Greek thought are quite remarkable and not easily transcended: exercising an inherent and distinguishing function makes something most fully what it is. In this manner the *being* of that something is most fully realized. That something, thereby, becomes most fully itself. In this way it most fully *is*. In most fully *being* it is *better* than it would have been, had it less fully exercised its function—and thus less fully been. To *be* is in itself good. *Fully to be* is best of all.

But consider. What happens if these Greek metaphysical assumptions are taken simply for what they are, *viz.*, assumptions. What happens, if the Greek identification of reality with permanence and, thus, the eternal, is itself put under renewed examination? Why then, for example, would something be good simply because it *was*, because it *is*? With respect to humans, if the linkage is broken between knowing and the (allegedly) eternal and unchanging nature of the known—a linkage whereby an immortality for the knower arises out of the eternality of the known—the goodness of what is, simply because it is, is surely undermined. Also undermined is any special status granted to transparent "knowing." And, however it is evaluated, this dismantling has in fact been the largely unarticulated philosophical trajectory of Western secular history.

Again, if the subtext motivating the epistemological linkage, *viz.*, that it provides a pathway to eternal life, is brought into question, all our standard assurances and value judgments are put in need of reconsideration. It could now be argued that what is, even and perhaps especially the agent of knowing, might more reasonably be subjected to restructuring rather than conservation, enhancement of capacity rather than the cultivation of longstanding attitudes of correspondence and obedience. If it is based upon being or becoming fully what it *is*, the goodness of what *is* is undermined without the resources of Greek metaphysics. If the triumph of nominalism is genuine, not only does the universal-particular distinction go, but with it a lynchpin of the doctrine of the inherent significance or sanctity of any and everything that *is*.

If we examine it even more closely, it has been the universal-particular distinction that has been taken to satisfy foundationally our *rational*[11] need for the unchanging. What is universal, unlike the particular, is purported to have in no literal or tangible sense spatial or temporal predicates.[12] In being instantiated rather than physically located, its status has been thought neither to be altered nor diminished by the vicissitudes of its varying, transitory instantiations, even by their extinctions. In being in one sense potentially anywhere at any time the universal is in another sense located at no specific "where" any of the time, thereby escaping the inevitable bruises involved in having a spatio-temporal location and an historical fate. The universal has had the status of the permanent. In its unchanging nature was found the condition of eternality. But in its sublime removal from concretely existing historical communities of human beings—and quite apart from the perplexing dynamics of its metaphysical appearances in time and its continuing dismissal by the advocates of nominalism—it has also lost concrete relevance and humanly binding force. For a variety of reasons the "universal" has become unfortunate and implausible baggage.

Again, by the arguments we have already traced, our (alleged) *knowledge* of the universal guarantees not so much its existence as strikingly exhibits something about ourselves. The existence of the universal, after all, is what makes our knowledge of it possible. The assumption that such knowledge is not only possible but in fact actual has underwritten the intuition, if not more fundamentally the *need* that we ourselves have to be eternal, the critical adjective for which ("immortal") arises out of the negation of our otherwise mortal status. If nominalism prevails, however, we undermine not only the universal but the notion of the unchanging. Either there turns out to be nothing unchanging, or it is not likely to be of a rational, i.e., universal nature. Correspondingly, if, *rationally* construed, there is nothing unchanging, nothing universal, then the presumption that reason is not only our core but both the source or our dignity and the locus of something immortal in us, is also undermined. But what is also thereby further undermined is any pretension to a foundation upon which a potentially universal community could be established. If such a community were ever to be engendered, it could not be brought into existence on the basis of any rational algorithm of choice, unless the very notion of reason were to be significantly transformed.[13] And what other sort of algorithm, what other means of overcoming incommensurabilities might be envisioned and subsequently sought than a *rational* one? This must become a most legitimate and pressing question. It cannot remain rhetorical.

However utopian, the ideal of commensurability and eventual agreement might still be sought, so long, that is, as a detached and uncontroversially exportable conception of rationality could be forged that would provide a plausible pathway to this ideal. A measure of leisure, disengagement and the capacity for sustained theoretical reflection would serve as its precarious preconditions. But if reason in this confidently ontological sense is discredited, the ideal of universally receivable, commonly accepted and humanly grounded Truth is made more fanciful than plausible. Though myriads of facts may be uncovered and recorded, once the universality of reason is discredited such facts cannot serve as premises for moral conclusions. Truth in any larger, rationally motivating, binding and integrating sense is simply unavailable. A community based on purportedly objective reality—things-in-themselves—that would found human togetherness and mutual interaction, thereby guaranteeing the possibility in principle of universal acceptance, becomes a chimera. Whether philosophically or scientifically "demonstrated"—surely the recurrent Jacobin delusion—such a community could neither resonate with concrete human actuality, nor minister to human need its in diversity and complexity.

With and after Hegel, of course, reason has in significant measure become what it can accomplish historically, if not through a dialectical development—spiritual, cultural or economic—then through its instrumentalization and technological application in various spheres of communication, commerce, and industry. Important to this post-Kantian thickening of reason—a development that brings the alleged autonomy of philosophy into question[14]—is an acknowledgement of what I have elsewhere called mediated reflexivity.[15] As humans we invariably relate in varying ways to ourselves, as opposed simply and unproblematically to being ourselves in any invariant substantial sense. Uncompromised ipseity is construed to be lacking, a deficiency captured without specific prejudice in the notion of reflexivity construed as self-relatedness.[16]

The relations humans sustain to themselves, however, essential as they are to their humanity, are themselves in turn not direct, but mediated. They arise within and in terms of the diverse and not always commensurable practices that have emerged within the context of historically based communities and their deeply embedded practices. In short, human awareness and understanding are situated (mediated), and their fabric is unavoidably woven into comprehensible and communicable patterns in terms of this situatedness. Though it is possible to express this circumstance within the vocabularies of the sociology of knowledge, the embeddedness of the human is deeper than the empirical resonance of sociological notions suggests. And a deep and ineradicable communal pluralism is an unavoidable consequence of this deep embeddedness.

I now turn to the grape. It adds a decidedly contemporary dimension to reflections on reason as a secular instrument for the construction of a universal community. Grapes are grapes. But they can also be enhanced and transmuted. Something can be done with them, the result of which in some cases is quite spectacular. The same may be true regarding our selves, including our rational dimension. It is evident that through biotechnologically and medically facilitated means it will become increasingly possible to enhance human performance and even human capacity itself with no likely neuro-physiologically negative side effects—not quite yet, of course, but as the century progresses.

On the secularly scientific view we have been tracing, it is hard to find compelling arguments that should deter those who seek competitive advantage through such means. What form might such an argument take? It will be painfully difficult to draft for those who hold that rights should primarily be construed as immunizations from interference rather than entitlements. And constraining arguments will be no less problematic when rights are construed as grounded in various voluntarily entered contractual agreements. What will function as the foundation(s) for constraints on various voluntarily gathered human beings? This problem is even worse to resolve for those who would construe rights as entitlements. How might we ground secular decisions with respect to various life transforming enhancements? Apart from abstractly formal considerations, in what will the rationality of the underlying decision procedures consist? Are we not returned to quandaries analogous to those we have already noted in exploring the notion of "things-in-themselves?" We are simply offered no plausible inference tickets, no algorithms for determining choices.

There is a further complication to be noted that may be unique to the twenty-first century. Incremental changes such as we are almost certainly going to see unfold through biotechnological and medical advances are likely to bring some of the more economically privileged to a "tipping point." A tipping point is where incremental changes suddenly give way to sometimes dramatic qualitative transformations.

I do not want to engage in technologically driven prophesy and am also mindful that the grapes to wine analogy is stretched when applied to human physiology. Yet, consider: brain chemistry does contribute to our (current) human nature. It contributes to and is part of what we human are "in ourselves." What are we supposed to do with respect to this chemistry? If we simply agree to let it be, not to violate it, we face at least two problems. First, evolution has not only made it what it is, but continues to do so. The expansion of our cerebral cortex over many millennia, for example, has been extraordinary. Is there a particularly strong reason to remain consciously passive in such matters just at the time that various interventions become possible that can accelerate and direct further evolutionary change? Second, the aggressive and territorial drives I mentioned before—correlated with the density of neural synapses in the hippocampus—are gaining even greater destructive potential through the advancing technological means of expression at their disposal. Is there any compelling reason to seek to contain these drives only at a very macro level?

Over the millennia we have seldom attempted simply to respect our evolving biological inheritance. We have sought means to control, overcome, or enhance it. By contemporary biotechnological standards, previous activities in this area—whether through deep meditation or drugs—have been largely unsystematic, episodic, and uninformed. But almost never have we simply left ourselves alone to be what we already were. In John Stuart Mill's sense of "experiments in living" there have almost always been such experiments—at least among those who have elicited our admiration or provoked our outrage. In the twenty-first century, a growing number of the affluent will engage in these experiments in living more intelligently and with the considerable support of medical science. Will various tipping points be reached regarding what we controversially call our human nature? For the first time in modern history this has become an open question. Secular answers to it are likely to be tangled and contradictory.

I now draw my remarks toward their conclusion. Most discussions of human nature have assumed a bifurcation between a higher and a lower dimension of that nature. The challenge has been to minimize and incorporate the lower into the higher. Over the long axial period of our history the higher has been correlated with the pure, disinterested, unchanging, and contemplative. The lower has been correlated with the instinctual, self-interested, changing, and active. Of course, there have been variations within this schema, but the schema itself has largely stood its ground. In its terms, part of our nature is to be encouraged. Supernatural help has often been sought for this purpose. At the same time we have sought to diminish other parts of our nature. The hope of some has been that these parts would be forgiven.

If one becomes post-metaphysically secular, however, the higher-lower distinction is quite hard to maintain. Even in a metaphysical age, it is problematic. On what grounds is the enduring and unchanging—by virtue of these traits alone—concluded to be of more value than the fleeting and changing? Is it not frequently the precarious and vulnerable that we most prize, seeking then to extend their longevity? Wittgenstein once asked, what problem would be solved by the supposed fact that we live forever. Well, we would not need to worry over what, if anything came "next" because there would be no next. There would be an infinite extension of what already is. There would be no worry over having to leave the theater before the show was over, for neither we nor the show would ever end. But Wittgenstein's challenge was surely a metaphysical one, directed by the surprisingly metaphysical Wittgenstein to the metaphysical in us: why the unchanging over the changing, eternity over time?

In a post-metaphysical world it will become increasingly difficult to translate the notion of higher into acceptable terms. Should it mean "more?" But of what? And for whom? To the second question the answer would seem to be: for ... everyone. Yet we shy away from this answer through various competing, if not often conflicting qualifiers: not for those who would take unfair competitive advantage of their dispensation, not for the undeserving, not if future generations are thereby disadvantaged, and so on.

Distribution issues aside, more of *what*? I will not rehearse the various answers to this question, for their rehearsal is one with what the histories of ethics and religion have offered and argued. We may contend with each other over which of these alternatives is more prescient or revelatory, but, rankings aside, most of us can agree regarding what these alternatives are.

Note in conclusion that I have spoken from a quite disengaged, spectatorial, though also provocative perspective. I have claimed (a) that whether in secularly oriented axial metaphysics or in contemporary science, there is no obvious reason why the way things are claimed to be should be taken as a guide to how we might choose to respond to them; (b) that from an evolutionarily informed, contemporary perspective there is much in how things are that is simultaneously unfortunate and productive—what could be called the alpha syndrome; and (c) that, seen against the backdrop of continually unfolding evolutionary change in any case, future biotechnological opportunities are likely in many cases to blur, if not render untenable the distinction between nature and artifice. In short, scientific advance may increase not just over longevity, but our plasticity and thereby render many previously inescapable aspects of ourselves optional. On such matters the jury must remain out.

Secularly established metaphysical notions of the sort I have explored offer little in the way of guidance toward our human future. A major reason for their failure is that they have provided no coherent foundation to support our past. Universality is a beguiling notion. It has deep, if ineffectual roots in our enduring human needs. A serious and open discussion of these needs must be undertaken at the intersection of religion and metaphysics. This intersection may well prove to be the unavoidable threshold our current time must cross.

Notes

1. I hesitate to employ the term "make" and thus put it in scare quotes. An outgrowth of the romantic reaction to Newtonian science, the modern distinction between creating (making) and discovering (finding) is dubious enough and has contributed a great deal of fuel to the idealist/realist controversy. This tangle of issues, however, is somewhat outside the scope of my current remarks.
2. On this admittedly provocative point, more needs to be said. Generally speaking, humanism is an import-driven position that tends to overlook its balance of trade and often even the fact of importation itself.
3. In referring to the human, I mean something more than merely biological constituents such as DNA composition, etc. The human must centrally involve the personal as it in fact does.
4. Some might say that reason equally qualifies. But this introduces a complex tangle of issues beyond the scope of this present undertaking.
5. Dostoevsky's most compelling account of this is found in *The Grand Inquisitor* scene of *The Brothers Karamazov*.
6. Here we encounter notions such as duties to oneself (Kant) or more general issues regarding self-relatedness. Especially after Hegel, these reflexive considerations introduce a considerable level of complexity that only invites confusion if not sorted out with extreme care.
7. We need look no further than Plato's *Phaedo* and its arguments for the immortality of the soul to appreciate the depth of these concerns and their intimate connections with each other. For an illuminating, if problematic analysis of the notions of universality and certainty in their historical impact on problems in the philosophy of mind see Richard Rorty's *Philosophy and the Mirror of Nature* (1979).
8. The full epistemic and ultimately reductive translation of this line of argument awaits the writing of someone like George Berkeley. If, following Aristotle as well, knower and known must be of the same genus, then mind could not know matter in any case. Given the widely adopted Newtonian claim that matter had no causal efficacy, what reason was then left to assume its existence at all? To have reached this point, however, was to have distanced the universal/particular distinction through the embrace of an already pervasive nominalism. What remained to take up the philosophical slack was the Cartesian problematic in which the mind/matter distinction became the driving force. It would not be too long before religious concerns got cast in the language of religious consciousness. As a consequence spiritual concerns became increasingly adjectival or adverbial in their expression rather than substantive, a circumstance we are yet fully to understand nor completely to escape.
9. I have given this notion an extended treatment in my *The (Coming) Age of Thresholding* (1999). The concept of *thresholding* itself emerges from reflections on the nature of the receptive dimension of our being.
10. It is Orthodox Christianity that has taken this assumption most seriously, though it finds recurrent, if embattled life within some of the mystical strains of Roman Catholic and Protestant faiths.
11. But how is one to understand rationality itself? The thicker the understanding the more culturally controversial reason becomes, the thinner the less relevant to actual human concerns. Over and beyond this consideration, the very notion of a rational *need* is hard to sort out without recourse either to platitudes or to a conception of human nature in which the very notion of reason has undergone transformation.
12. Perhaps it was this seemingly unbridgeable chasm more than anything that exorcized Plato's most influential, subtle and revisionist student.
13. That it could be transformed in the appropriate religious context I have no doubt.
14. What is offered is either the abstract *a priori* of Kant or a seemingly *a posteriori* web of empirical practices. If the latter one as is, where has the supremacy of philosophy gone?

15. See *Language and Being: An Analytic Phenomenology* (1970) and *The (Coming) Age of Thresholding* (1999).
16. Out of this come the puzzling notions of identity and difference, puzzling at least to analytic philosophy.

Bibliography

Erickson, Stephen A. (1970). *Language and being: An analytic phenomenology.* Yale University Press: New Haven and London.
Erickson, Stephen A. (1999). *The (coming) age of thresholding.* Kluwer Academic Publishers: Dordrecht.
Rorty, Richard (1979). *Philosophy in the mirror of nature.* Princeton University Press: Princeton.

Nature as Second Nature: Plasticity and Habit

Peter Wake

> *For it is the nature of humanity to press onward to agreement with others; human nature only exists in an achieved community of minds.*
>
> (Hegel, 1977, § 69, translation altered)

I will begin where Stephen Erickson ends "Reflections on Secular Foundationalism and Our Human Future," with the grape (Erickson, 2009). His speculations concerning the moral issues that will likely arise in light of the development of new biotechnologies draws into question the very possibility of clearly dividing human nature from cultural artifice. They draw into question the possibility of distinguishing those traits which differentiate us from the beasts and underwrite our immortality from those which are malleable and impermanent. Nietzsche claims, famously, at the outset of *The Birth of Tragedy* that the pre-Platonic Greeks described the origin of the work of art through the coupling of Apollo, the god of dreaming, among other things, and Dionysus, the god of the fermented grape. The creation of a work of art, or—Nietzsche would surely add—a philosophical theory of human nature, first requires a distancing or separation from the world. Through the experiences of dreaming and intoxication—intoxication as the effect of the grape—we are, at a primordial level, afforded this distance: the beings of the world are not simply *there*; what is immediately given is not what must be; instead, through these experiences, one sees that the world, and oneself, could be other than what they are. The god of the fermented grape represents then a "drive" (*Trieb*) that is natural and unnatural at once (Nietzsche, 1993, p. 14). It is a drive that arises from *within* nature itself, yet, at the same time, pushes beyond nature to what it is not—call this "culture," "art," "biotechnology." The product of this drive is what Nietzsche describes as "a *monstrous* opposition" (*ein ungeheurer Gegensatz*) (1993, p. 14). This emergence from within nature of the division *between* nature and human self-consciousness is an aberration, a monstrosity, both natural and unnatural as once.[1] As such, this drive is akin to what Hegel calls the "tremendous power [*ungeheure Macht*] of the negative" (1977, § 32): the power of the "pure I" to negate what is. According to Hegel, this is the power of the Understanding (*Verstand*), "what was earlier called the Subject" (§ 32), and it works to interrupt immediate, self-enclosed and unreflective Being. From the theoretical perspective, this original distancing—this power

P. Wake (✉)
Department of Philosophy, St. Edward's University, Austin, TX, USA
e-mail: peteraw@stedwards.edu

M.J. Cherry (ed.), *The Normativity of the Natural*, Philosophical Studies in Contemporary Culture 16, DOI 10.1007/978-90-481-2301-8_10,
© Springer Science+Business Media B.V. 2009

of the negative—allows us to reflect upon ourselves, to be a question to ourselves. Practically, it means we are a work in progress, we are underway, we are free.

Despite their many differences, both Hegel and Nietzsche present the division between nature and culture as arising from within nature itself and as such, they are both forced to confront the difficulty of rigorously distinguishing human nature from human artifice. In other words, they are forced to confront the type of question that Erickson's conclusion invites: Can we transmute and enhance ourselves through human artifice in a similar manner to the way in which the grape can be enhanced and transmuted into wine? Is it the case that our nature is to be *other* than our nature? Is it *malleable* and thus something to be created? Hegel and Nietzsche's response to these kinds of questions is shaped by their recognition of what we might call, following Catherine Malabou, the *plasticity* of human nature.[2] "Plasticity," *Plaztizität*, is derived from the Greek *plassein*, to mold, model, or form, and the adjective "plastic," *plastisch*, has the following double meaning: "on the one hand, to be 'susceptible to changes of form' or malleable (clay is a 'plastic' material); and on the other, 'having the power to bestow form, the power to mould,' as in the expression 'plastic surgeon' and 'plastic arts.'" And while what is "plastic" is not "rigid" or "ossified," it must also be distinguished from the "polymorphous": "things that are plastic preserve their shape, as does the marble in a statue: once given a configuration, it is unable to recover its initial form."[3] Building upon the double and opposed meaning of "plastic" as what is "at once capable of giving and receiving form," Malabou argues that "plasticity" can be conceived as the "name for the originary unity of acting and being acted upon, of spontaneity and receptivity" (2005, p. 186). When applied to human nature, it names the very process of self-determination itself, for as we will see, this unity of spontaneity and receptivity translates into the properly speculative, Hegelian idea of freedom as the synthesis of freedom and necessity. At this point, we need only add that according to this plastic conception of human nature it is clearly not to be understood on the model of a substance with one distinct quality—whether this be consciousness, language, reason, or freedom—that distinguishes the substance we are from all other substances. Appealing to *plasticity* in order to characterize human nature does not simply repeat this gesture of isolating and privileging a particular quality because plasticity is not a quality attached to a pre-existing substance. The truth of human nature as plasticity must be conceived, in Hegel's words, "not as *Substance*, but equally as *Subject*" (1977, § 18, translation altered). As such, the plasticity of the subject describes the idea that the form—the "nature"—of this subject is not something that is settled in advance.[4]

In what follows, I will turn briefly to Nietzsche to help clarify the idea of the plasticity before returning to Hegel to consider his attempt to overcome the opposition between human nature and artifice.

*

One of Nietzsche's guiding philosophical projects was to "overturn" Platonism and the correlative faith in the "real" world of things-in-themselves.[5] Erickson avoids this dramatic language, but at the heart of his critique of secular reason is a similar challenge to the core metaphysical assumptions of Platonism—or, at the very least, the ability of secular reason to defend these assumptions. Erickson prefaces his reflections on the limits of secularism by drawing attention to moments of conflict between philosophical orthodoxy and the lived experience of the time. Thus, Schoolmen were confronted by the early figures of philosophical modernity while the Idealism of Bradley and McTaggert was challenged by Russell and Wittgenstein. Surely, one of the most decisive historical examples of this kind was the quarrel (*diaphora*) between Plato and the tragic poets that we find at the inception of the Platonic tradition itself.[6] Against the malleability of human nature that Nietzsche finds at the heart of the preplatonic "tragic age of the Greeks," and the difficulty in strictly distinguishing nature from culture that this introduces, we find in Plato, and the history of Platonism generally, the quest for permanence. What this quest means when applied to human nature is that to determine the *nature* of human beings is to discover what is unchanging and thus immortal in us. Once the quality that defines this nature has been uncovered, it can then provide the criterion for how we ought to act and Erickson argues that this quest for permanence and certainty, along with the supposition that our actions ought to conform to these permanent, universally human traits, remains the often-unacknowledged foundation of philosophical secularism.

Erickson names Hegel and not Nietzsche as the one who untangles the complexities inherent in the Platonic answer to what makes human nature universal and thus a reliable guide for our actions. We are told that in an untangled Platonism what distinguishes human beings from animals is our *capacity* for knowing what is permanent. Fulfilling this capacity makes the knower increasingly like the known, which is to say, permanent. This transformation of the knower through the activity of knowing itself presumes a certain malleability inherent in the nature of the knower. As Erickson writes, "knowing is that which enables us to *achieve* our true nature" (2009, p. 130). Human nature, then, entails the achievement of what we already are. Or, in Erickson's words, "exercising an inherent and distinguishing function [here, knowing what is permanent] makes something most fully what it is" (p. 131). We find this ontological argument in a very concise form in Book I of the *Nicomachean Ethics*. The proper function (*ergon*) of a being is determined by what distinguishes it from other beings (2002, 1097b25). Plant life consists in "taking in nutriment and growing" (2002, 1098a2), but it does not include perception. Animal life requires nutriment and growth, as well as perception, yet it does not possess the capacity to reason actively (2002, 1098a7). Human beings do actively reason, but they are not immortal. Given this, the *ergon* of human beings is active reasoning. We achieve our humanity when we do this well, and yet, at the same time, to realize this potential is to be *more* than human. In Book X of the *Ethics*, when defending the philosophical life of reflection over the practical life of the statesman, Aristotle claims that "such a life will be higher than the human plane; for it is not in so far as he is human that he will live this, but in so far as there is something divine in him..." (2002, 1177b26–28). Thus we have the Aristotelean formulation of Erickson's claim that the knower has the potential to become increasing like that which is permanent, and so immortal.

Although Hegel lays bare this logic, Erickson claims that he nevertheless remains within its grip. Unlike Nietzsche, he neither aspires to "overturn" Platonism, nor does he reject the possibility of knowing things-in-and-for-themselves. Hegel holds that through active, rational knowing, we achieve our human nature, yet what is achieved is, in Erickson's words, "necessarily already in place" (2009, p. 130). According to this interpretation of Hegel, it has taken the history of human spirit to make manifest what we have always already been—that which is divine in us—and this concrete manifestation is, in effect, the apotheosis of human history itself. Hegel's System, in turn, provides the philosophical defense of this apotheosis. Yet far from history being the dialectical synthesis of the finite knower and the infinite object of knowledge, Erickson claims that the Platonic identification of reality with what is permanent has increasingly been revealed as a metaphysical assumption and, following from this, the "unarticulated" philosophical trajectory of western secularism has been the increasing *rupture* between the knower and this infinite object (p. 131). Given that the union of the knower and the known has been the ground for claims by human beings to participate in what is immortal, Erickson's analysis invites the question: What are we left with when this link becomes severed?

One response to this rupture is that at the end of the secular philosophical tradition, with its deep Platonic roots, we come to recognize precisely what Nietzsche discovers in the pre-Platonic tragic conception of the self, namely, the utter malleability of human nature. This is not the malleability that allows for the achievement of what we have always in essence been. It is not, in other words, the position ascribed above to Hegel. Rather, with the loss of any right to claim that there is such an unchanging essence, we are simply left with *malleability* per se and the view that human nature would be best understood on the model of a work of art.[7] Faced with the inability of the secular philosophical tradition to defend a universally valid notion of human nature, Nietzsche rejoices.[8] In the conclusion of "How the 'Real World' at Last Became a Myth," Nietzsche's concise account of the rise and decline of Platonism, he exalts the expiration of the idea of the "real," unchanging world that is thought to stand in marked opposition to the world of mere appearance: "We have abolished the real world: what world is left? The apparent world perhaps? . . . But no! *with the real world we have also abolished the apparent world!* (Mid-day; moment of the shortest shadow; end of the longest error; zenith of mankind . . ." (1990, p. 51). A self-proclaimed enemy of nihilism, he interprets the abolition of the fable of the thing-in-itself as the potential for new life. Erickson also frames his challenge to secularism in terms of revitalization, but he does so in a way that, on the surface, seems quite foreign to Nietzsche's thought, for he writes that this revitalization can take place through the transformation of the notion of reason within the "appropriate religious context" (p. 136). What would this transformed and resuscitated reason look like? We do know from Erickson's reflections on secular humanism something about what it is *not.* As we have seen, he argues that an unembellished secularism claims to have the capacity to know human nature as it is in itself and it holds that our knowledge of this thing-in-itself ought to serve as a guide for our actions. Yet, even if secular reason were able to know this (a position Erickson rejects), it would be unable to justify the claim that any object, including persons, has inherent value. Instead, it can only arrange beings according to a logic of "support." Secular reason might follow Aristotle in his *Politics* and argue that there is a natural hierarchy such that the purpose of plants is to be eaten by animals and the purpose of animals is to fatten human beings.[9] But despite its desire to do otherwise, it would be unable to challenge Aristotle when he goes on to claim that war, as a natural art of acquisition, exists so that those stubborn people who are by nature meant to be governed, but refuse to submit, are returned to their proper position of enslavement.[10] Any claim to the inherent value of human beings would transcend secular reason itself. To the extent that it shies away from this conclusion and asserts what Kant calls the "unconditional, incomparable absolute worth" of the person (1998, p. 43),[11] it represses the fact that such assertions are parasitic upon what it is not—that is, as Erickson formulates it, reason placed within the "appropriate religious context." There is, relative to its own terms, an unacknowledged leap of faith at the heart of secular humanism. Put more starkly, it cannot ground claims to our potential for rational self-determination; human autonomy is not self-grounding.

What, then, would a resuscitated reason look like positively? If, as Erickson argues, secular reason is unable to defend the absolute worth of all persons, would reason, properly transformed, aim to do so? We are told that this transformed reason would account for the essentially hermeneutical nature of knowing, which is to say, it would recognize that all knowledge is mediated by, or situated in, a concrete historical context. This is consistent with Erickson's various appeals to the value of the particular and the unique, as opposed to the tradition valorization of the universal (p. 132). If the severing of the universal from the particular—the pure moral law from what is to be done in a concrete situation—is the "largely unarticulated philosophical trajectory of Western secular history" (Erickson, p. 131), would a resuscitated reason attempt to bridge this gap by way of a genuinely *concrete* universal? Furthermore, if burgeoning technological advances will inevitably strain secular attempts to maintain the distinction between nature and culture, would a transformed reason strive to keep these spheres separate?

These questions extend beyond the scope of Erickson's paper. As a way of coming to grips with them, however, I propose developing a sketch of a possible Hegelian response. The decision to turn to Hegel is based on certain core positions he and Erickson share. Whether or not a commitment to the plasticity of human nature is one of these is a question I raise without presuming to provide a response.

*

Erickson's challenge to the formalism of Kant's moral philosophy and his insistence that knowledge, including what is discovered through practical reasoning, must be mediated by historical context are consonant with Hegel's thought. Hegel's concept of "ethical life" (*Sittlichkeit*) should be understood, in part, as a reaction to the limitations of the sublime dictates of Kantian practical reason.[12] With Erickson, Hegel holds that the quest for universality comes with the sacrifice of the particular or unique—"the concretely existing historical communities" (p. 132)—and as such, it undermines the capacity of ethical laws to guide our actions. Furthermore, Erickson, like Hegel, recognizes a malleability in human beings that poses a significant challenge to a meaningful separation of human nature and artifice. But in so far as Erickson understands Hegel to be a part of the Platonic tradition, Hegel would presumably not follow Nietzsche in reducing human nature to malleability per se and championing the *aesthetic* imperative of self-creation.[13] Instead, for Hegel, the historical unfolding of human nature achieves what was, in Erickson's words, "necessarily already in place." But what exactly is it that is "already in place"?

Despite Hegel's unquestionable indebtedness to Aristotle (Emil Fackenheim goes so far as to say that the "Hegelian system is a modern restatement of Aristotelianism," 1967, p. 179n), we misunderstand Hegel's thought if we think of human nature—what is "necessarily already in place"—as that quality, or function, which distinguishes us from plants, animals, and God. If the error made by secular reason is its conviction that discovering the human thing-in-itself will aid us in guiding human conduct, it is rooted in a misguided conception of ourselves as a substance, as a "rational animal" or a "thinking thing." Rather than a substance with a nature defined by its essential qualities, Hegel argues that the self should be conceived in terms of its *concrete freedom*, a freedom that is, as Hegel says, identical with necessity. As he writes in the *Encyclopaedia Logic*, it is

> absurd ... to regard freedom and necessity as mutually exclusive. To be sure, necessity as such is not yet freedom; but freedom presupposes necessity and contains it sublated within itself. The ethical person is conscious of the content of his action as something necessary, something that is valid in and for itself; and this consciousness is so far from diminishing his freedom, that, on the contrary, it is only through this consciousness that his abstract freedom [a potential to be free] becomes a freedom that is actual and rich in content, as distinct from freedom of choice, a freedom that still lacks content and is merely possible (1991, § 158, Addition).

We can understand this to mean that far from a moral law being something that is "necessary" and that, as such, impinges upon our (arbitrary) freedom to simply do what we please, our actions are free only insofar as we take up this law as our own. Conversely, we could also take our *desires* as what are "necessary" and thus as what interferes with our formal freedom to abstract from desire and instinct as such. Hegel describes this formal freedom as the capacity to "withdraw itself from everything external and from its own externality, its very existence; ... in other words, it can keep itself affirmative in this negativity and possess its own identity" (1971, § 382). But, of course, formal freedom as the mere capacity to abstract from existence, along with the identity associated with it, is completely empty. For Hegel, absolute freedom is concrete, which is to say, it "is not a possibility or capacity but a determinate way of acting" (Wood, 1990, p. 39). It is to be with oneself in one's desires. Freedom made concrete is to desire to do what one ought to do.

But whether we conceive of freedom as formal, negative, civil, subjective, concrete or absolute, the question remains, how is it different from the other qualities that have been isolated to distinguish human nature from the nature of other beings? As I indicated above, Hegel's response is that the freedom defining the self is not a quality that characterizes a pre-existing substance or nature, but *negation*. The self as negation has been described concisely in this way,

> [T]he subject is in no way the *self all to itself*. It is, to the contrary, and it is essentially what (or the one who) dissolves all substance—every instance already given, supposed first or

> last, founding or final, capable of coming to rest in itself and taking undivided enjoyment in its mastery and property (Nancy, 2002, p. 5).

In the *Phenomenology of Spirit*, Hegel himself writes, apropos of the slave who has been "seized with *Angst*" before death: "[T]his pure universal movement, the absolute melting-away of everything stable, is the simple, essential nature of self-consciousness, absolute negativity, pure being-for-self, which consequently is implicity in this consciousness" (1977, § 194). As far as the self is concerned, there is no *thing*-in-itself, no nature to be discovered. We *are* nothing more than freedom in the practical domain and, in the theoretical domain, self-reflexivity.

As the dialectical synthesis of freedom and necessity, absolute freedom is best conceived in terms of plasticity, and not substance, for plasticity is precisely the "unity of acting and being acted upon, of spontaneity and receptivity" (Malabou, 2005, p. 186). We would not be who we are but for the language we speak, the traditions, customs, and rituals we have inherited, the world of meaning into which we have been thrown. At the same time, it is only against this backdrop that we are able freely and concretely to act. Beyond this, our actions, understood collectively, maintain and transform this mobile, plastic backdrop. In other words, absolute, concrete freedom is not something that stands in opposition to what it acts upon; instead, it is *inseparable* from each case in which it is determined. Freedom, as our "nature," is not distinct from what it does or produces. Each term in the binary opposition between nature and culture stands as a one-sided abstraction from this original, living unity. Ultimately, the problem of how to distinguish what is truly of our nature and what it merely accidental is a false one. As Hegel claims, "the tremendous power of the negative," the absolute power of thought, can take what is accidental and enable it to "attain a separate existence of its own and a separate freedom" (1977, § 32). It is able, through recollection, to make what is initially accidental into something that is essential. The essential—human nature—does not, for Hegel, exist as "necessarily already in place." It is produced in the sense that the *criterion* by which an essence is judged to be an essence in the first place has a history. Indeed, for Hegel, the conflicts that arise between our different criteria of truth and our experience is precisely what fuels the unfolding of the history of consciousness itself. Thus, he writes in the "Introduction" to the *Phenomenology*, when confronting the Kantian problem of the possibility of knowing the thing-in-itself,

> an examination consists in applying an accepted standard, and in determining whether something is right or wrong on the basis of the resulting agreement or disagreement of the thing examined; thus the standard (and Science likewise if it were the criterion) is accepted as the *essence* or as the [thing-] *in-itself* (1977, § 81).

Yet, for Hegel, "[c]onsciousness provides its own criterion from within itself" (1977, § 84). What stands as the essential thing-in-itself cannot be dissociated from this (accidental, non-essential) origin.

*

As we have seen, Aristotle develops his thought on human nature from within a metaphysics of substance and as such, the criterion by which the essence of human nature is determined is deemed to be its distinct *ergon* or function. On this point, Hegel diverges from Aristotle's thought. Aristotelean substance is not equally subject (1977, § 17). Nevertheless, Aristotle's analysis of habit (*schēma*) and disposition (*hexis*) is exemplary of how to conceive of the plasticity of human nature. Taking his lead from Aristotle's *De Anima*,[14] Hegel strives to show how habits work to translate the psychic and somatic into one another. Body and soul mutually fashion each other and through this process, they develop into what they are and have never previously been (Malabou, 2005, p. 26). Hegel's relation to Aristotle in this regard reflects his overarching aspiration to synthesize Greek and Modern thought. What he affirms in Aristotle's analysis of habits is precisely the process whereby an accident is made essential. As Malabou argues, this is paradigmatic of Greek thought. Modern thinking moves in the other direction: the essential is made accidental, and the model for this is the *kenosis* of God the Father into Jesus the Son (p. 82). In so far as Hegel rejects the substance metaphysics of Aristotle, he is a Modern. In so far as he affirms the centrality of plasticity and habit in human nature, he remains a Greek.

"Habit is rightly called a second nature" (Hegel, 1971, § 410). According to Hegel, it is a *nature* in that it is something substantial. It exists stably over time. It subsists. Furthermore, habit is a kind of immediacy: "it is an immediate being of the soul" (1971, § 410). But it is a *second* nature precisely because this immediacy has been posited *by the soul*. Habits are the products of intervention, formation, and cultivation. But if there is a second nature, what is our "first" nature? What is it that we *immediately* are? Hegel's position is, in effect, that claims to natural immediacy are always retroactively determined, and thus mediated. That is, thinking in terms of an "immediate nature" is always an abstract and one-sided determination. Philosophy as a retrospective activity, as recollection,[15] must recognize that the return to the origin—what is "already in place"—is the return to a ground that is no longer immediate, that is no longer "natural," and thus it is not, properly speaking, a "return" at all. As mediated, then, the concept of an original human nature—whether this origin is conceived historically or logically—involves a retroactive construction.

As that which allows human nature to *become* a nature, habit, in effect, operates according to the logic of sublation (*aufheben*). When defining this concept, Hegel speaks of the uncanny existence of two opposed meanings within one word:

> *[Aufheben]* means to preserve, to maintain, and equally it also means to cause to cease, to put an end to. Even 'to preserve' includes a negative element, namely, that *something is removed from its immediacy* [my italics] and so from an existence which is open to external influences, in order to preserve it. Thus what is sublated [*das Aufgehobene*] is at the same time preserved; it had only lost its immediacy but is not on that account annihilated (1969, p. 107).

At the broadest level, *aufheben* presents a logic of movement. More specifically, it attempts to describe the movement of qualitative change. More specifically still, sublation accomplishes a union of apparent opposites and, as such, not only does it both *cancel* and *preserve*, but it also *lifts* or *raises up*. Most importantly, for our purposes, *aufheben* is itself subject to this process, which is to say, *aufheben* has a history. Presenting this process in its pure, logical form, as Hegel does in his philosophy, *is* the *Aufhebung* of prior manifestations of the movement of *aufheben*. According to Malabou, in the Greek world this movement is found in Aristotle's analysis of habit.[16] As a concrete manifestation of this logic, habit is that which allows the subject—*avant la lettre*—to develop into a subject. When a practice becomes habitual, it stagnates, loses its living meaning and ossifies.[17] At the same time, it is only because of the stability provided by habits that we can act in a determinate way. Only against the background of an entrenched set of habits can free activity avoid being merely an abstract potentiality.

It is for this reason that we can say that the body pervaded by the soul *is* freedom.[18] Habits can be freely formed and molded while being, at the same time, the inheritance that is necessary for free, determinate action to occur in the first place. For free action arises only when the body no longer stands as a foreign and opposed entity, but has been formed, as Hegel writes, "into a pliant and skilful instrument of its [the soul's] activity." By transforming the body in this way, the "soul relates itself *to itself* and its body becomes an accident brought into harmony . . . with freedom" (1971, § 410, Addition, my italic). The soul knows itself as a soul only through the body, and the body transcends its status as an accident by way of the formative activity of the soul. Through the repetitions that develop into habits, there eventually "arises a *magical* relation, an immediate operation of mind on body" (§ 410, Addition). Against the traditional Platonic idea of human nature, human beings do not have a substance; the subject nevertheless exists, but it exists as a process. And through this process, what is accidental is made essential. We might think here of Aristotle's claim that "we become . . . courageous by doing courageous things" (2002, 1103b1–2). In order to become a truly courageous person, one creates the disposition, or *hexis*, of a courageous person by, in part, repeating the acts that the *polis* deems to be courageous. Or, to use an example that Hegel provides, when we learn to write, the activity of writing eventually becomes a habit such that we have "so completely mastered all the relevant details, have so infected them with its universality, that they are no longer present to us as single details and we keep in view only their universal aspect" (§ 410, Addition). When this occurs, there is the immediate, "magical" union of soul and body, thought and the written word, but, of course, this union is mediated by the repetition that has allowed us to master these skills in the first place. As such, habits are spectral in character. They mediate between body and soul, yet, at the same time, the effect of this mediation is that "the soul, on the one hand, completely *pervades* its bodily activities and, on the other hand, *deserts* them, thus giving them the shape of something *mechanical*, of a merely *natural effect*" (§ 410, Addition).[19] The character of habit is to be both present and absent at once. Writing, as a habit, is present and pervades every stroke of the pen, but at the same time, as habitual, the activity is mechanical, thoughtless, and thus also absent.

Human nature as plastic has this same character of being present and absent at once. Far from human nature being the actualizing or "making present" of what was already in place as a potentiality, it is instead like the rhythm that hovers between and unites metre and accent. Rhythm as the sublation of metre and accent: this is the analogy Hegel uses to describe the union of the subject and predicate in a properly philosophical—that is, *speculative*—proposition. With this, we turn from the act of writing to the production of meaning itself. In the "Preface" to the *Phenomenology*, Hegel claims that the philosophical proposition alone is capable of achieving the "goal of plasticity" (1977, § 64). As typically understood, a proposition reproduces the logic of substance: the subject of a proposition is thought of as a static center that supports an array of distinct predicates in the way that a fixed and stable substance supports any number of individual determinations or accidents. In the propositional judgement, "the ink is black," *ink* is distinguished from other possible substance-subjects by the accident-predicate *black*. By contrast, reading a philosophical proposition—"God is Being" (1977, § 62)—unfolds in such a way that the subject, as a passive and inert ground, is unmoored. To move from "God" to "Being" is not simply to attribute a predicate to subject. The speculative proposition does "not 'predicate' something newly discovered about the stable subject that was there already" (Harris, 1997, p. 143). For Hegel, any philosophical attempt to conceive of God will always begin with a prior, inherited notion of what God is, but when confronted by the predicate, "Being," this act of predication *changes* the subject. It is forced back upon the subject in order to "remodel" it and this further development of the concept is possible only because of its plasticity.

Hegel describes the meaning of the properly philosophical proposition as the "floating center"—the circling back and forth—between subject and predicate (1977, § 61). What is proper to the human being is the floating center—the rhythm—that unites the movement between nature and artifice. With human nature, conceived in terms of plasticity, there is nothing that is already there, for self-actualization is not a goal that can be predetermined. Freedom can only be determined by way of actions done in the name of freedom. As we have seen, from the perspective of the metaphysics of substance, the historical manifestation of freedom will be understood as the actualization of an inherent potentiality. In so far as we are marked by the spiritual need for the kind of mutual recognition that can only be "achieved in a community of minds" (1977, § 69), is this not the fulfillment of a need that is "already in place"? No. Any conception of what is already in place is itself retroactively produced. If we discover that we are those beings who desire to be able to say, "I am I," and, as Hegel argues, we find that self-recognition can only takes place through another, this does not define a pre-existing potentiality in need of actualization. If, as Hegel writes in the *Phenomenology*, "it is the nature of humanity to press onward to agreement with others" (§ 69), this is because an individual self-consciousness only exists *as* self-conscious in relation to another. Again, it does so not because it is in its pre-existing nature to do so, but because it becomes itself only in relation to another. Positing it as an original potentiality is just that, an act of positing an origin. Human nature is always a second nature.

Notes

1. That this monstrous quality touches self-consciousness and the drive to intelligibility that is found, in its most excessive form, in the philosopher is evident from Nietzsche's description of the

critical, rather than *creative*, instinct that he finds in Socrates as "a monstrosity *per defectum*!" (Nietzsche, 1993, p. 66). John Sallis describes the defect in this way, "What is monstrous is the lack of the natural relation between instinct and conscious knowledge, that is, the divergence from nature lies in the inversion" (Sallis, 1991, p. 124).

2. See Catherine Malabou, *The Future of Hegel: Plasticity, Temporality and Dialectic* (2005). I am indebted to Malabou for her study. She not only provides an extensive account of the role that *plasticity* plays in Hegel's *Philosophy of Spirit*, but she also presents a persuasive case for the central, if largely unacknowledged, position that it holds in his philosophy as a whole.
3. "'Plastic', thus, designates those things that lend themselves to being formed while *resisting* deformation" (Malabou, 2005, pp. 8–9).
4. I take this description of plasticity as a form not settled in advance from John Russon.
5. Heidegger quotes the following observation that Nietzsche makes in one of his early notes (1870–1871), "My philosophy an inverted Platonism: the farthest removed from true being, the purer, the finer, the better it is. Living in semblance as goal." Heidegger makes the following remark in response to this observation: "That is an astonishing preview in the thinker of his entire later philosophical position. For during the last years of his creative life he labors at nothing else than the overturning of Platonism" (Heidegger, 1979, p. 154).
6. *Republic* X, 607b. For the quarrel between philosophy and poetry, see also *Laws* 967c–d.
7. At the culmination of the *Encyclopedia Anthropology*, Hegel himself defines "man" as "the soul's work of art" and, as Malabou notes, he does so in direct reference to Greek sculpture (Hegel, 1971, § 471; see Malabou, 2005, p. 68).
8. In *The Birth of Tragedy*, Nietzsche writes that "now ... science is rushing irresistibly to its limits, where the optimism essential to logic collapses. For the periphery of the circle of science has an infinite number of points, and while it is as yet impossible to tell how the circle could ever be fully measured, the noble, gifted man, even before the mid-course of his life, inevitably reaches that peripheral boundry, where he finds himself staring into the ineffable. If he sees here, to his dismay, how logic twists around itself and finally bites itself in the tail, there dawns a new form of knowledge, tragic knowledge, which needs art as both protection and remedy, if we are to bear it (Nietzsche, 1993, pp. 74–75).
9. "We may infer that, after the birth of animals, plants exist for their sake, and that the other animals exist for the sake of man" (Aristotle, 1988, 1256b15–18).
10. "[T]he art of war is a natural art of acquisition, for the art of acquisition includes hunting, an art which we ought to practice against wild beasts, and against men who, though intended by nature to be governed, will not submit; for war of such a kind is naturally just" (Aristotle, 1988, 1256b21–26).
11. Erickson notes that it is "against the backdrop of Kant's enlightenment rationalism ... that the secular project has made most of its claims to progress" (p. 129).
12. We can say very generally that "ethical life," *Sittlichkeit*, corresponds to the public morality that encompasses the communal life of a "people" (Hegel, 1975, p. 92). Indeed, Hegel presents *Sittlichkeit* as the German equivalent of the Greek *ethos* (p. 112). He refers to it as "substance" in order to draw attention to its stability, and to distinguish it from the negativity of subjectivity, but, unlike nature, this substance is born of free action and its laws are self-imposed (see Allen Wood, *Hegel's Ethical Thought*, 1990, p. 198). In contradistinction to *Sittlichkeit*, *Moralität* denotes the modern moral thinking exemplified by Kantian moral theory that emphasizes individual, self-reflexive deliberation.
13. "... the existence of the world is *justified* only as an aesthetic phenomenon" (Nietzsche, 1993, p. 8). Gilles Deleuze writes of Nietzsche's doctrine of will and freedom: "*the will to power is essentially creative and giving*: it does not aspire, it does not seek, it does not desire, above all it does not desire power. It gives ..." (Deleuze, 1983, p. 85).
14. In the *Encyclopaedia Philosophy of Mind*, Hegel's presentation of the development of soul is fundamentally oriented by Aristotle's *De Anima*: "The books of Aristotle on the Soul ... are still by far the most admirable, perhaps even the sole, work of philosophical value on this topic" (see Hegel, 1971, § 378; cited in Malabou, 2005, p. 25).

15. At the conclusion of the *Encyclopaedia*, Hegel addresses what he argues is the highest form of spirit, philosophy, and he defines it as a movement that "only looks back on its knowledge" (1971, § 573). It is a return "back to its beginning" (§ 574).
16. On the history of the *Aufhebung*, see Malabou, 2005, pp. 144–145.
17. Comprehending the process by which meaningful actions and practices become ossified into dead rituals has been one of Hegel's most long-standing concerns. In an essay from 1795 to 1796 entitled "The Positivity of the Christian Religion," Hegel gives the example of an annual rite, still practiced in his own time, of the prelate washing the feet of the poor. He locates the origin of the ritual in the Jewish tradition of hospitality where washing the feet of a guest was a daily activity and a courtesy to guests. The contemporary continuation of the ritual is clearly meant to affirm the principles of equality and humility by illustrating the transience of the temporal power and luxury that the prelate enjoys, but the rite becomes a "comedy" because the principles are realized in the symbolic realm of the prelate's gesture and nowhere else (Hegel, 1971, p. 89). The source of this comedy is a society that needs, according to its own stated principles, to reconcile the extremes that arise within it, but the attempted reconciliation involves two dissymmetrical sacrifices: the enduring material sacrifice of the poor and the symbolic, theatrical gesture of sacrifice made by the prelate. The ritual is comic in so far as it has the effect of leaving everything as it was. But beyond what Hegel sees as the hypocrisy of the practice, the loss of the "living" character of ritual stems, in part, from the fact that washing the feet of the poor has no relation to what people actually do in late eighteenth century Germany. The source of the practice, the tradition of hospitality, has been lost.
18. In habit, "man relates himself not to a contingent single sensation, idea, appetite, etc., but to himself, to a universal mode of action which constitutes his individuality, which is posited by himself and has become his own and for that very reason appears as free" (Hegel, 1971, § 410, Addition).
19. With passages like this in mind, we can see why Malabou would write of the "virtual" character of habits (2005, p. 55).

Bibliography

Aristotle (1988). *Politics*, S. Everson (ed.). New York: Cambridge University Press.

Aristotle (2002). *Nicomachean ethics*, S. Broadie and C. Rowe (trans.). Oxford: Oxford University Press.

Deleuze, G. (1983). *Nietzsche and philosophy*, J. Thomlinson (trans.). New York: Columbia University Press.

Erickson, S.A. (2009). Reflections on secular foundationalism and our human future, in *The normativity of the natural*, M.J. Cherry (ed.). Dordrecht: Springer.

Fackenheim, E. (1967). *The religious dimension in Hegel's thought.* Bloomington: Indiana University Press.

Harris, H.S. (1997). *Hegel's ladder I: The pilgrimage of reason.* Indianapolis: Hackett Publishing Co.

Hegel, G.W.F. (1969). *Science of logic*, A.V. Miller (trans.). Atlantic Heights: Humanities Press International.

Hegel, G.W.F. (1971a). *Philosophy of mind. Part 3 of The encyclopedia of philosophical science*, J.N. Findlay (trans.). Oxford: Oxford University Press.

Hegel, G.W.F. (1971b). The positivity of the Christian religion, in *Early theological writings*, T.M. Knox (trans.). Philadelphia: University of Pennsylvania Press.

Hegel, G.W.F. (1975). *Natural law: The scientific ways of treating natural law*, T.M. Knox (trans.). Philadelphia: University of Pennsylvania Press.

Hegel, G.W.F. (1977). *Phenomenology of spirit*, A.V. Miller (trans.) Oxford: Oxford University Press.

Hegel, G.W.F. (1991).*The encyclopaedia logic. Part I of The encyclopaedia of philosophical sciences*, T.F. Geraets, W.A. Suchting, H.S. Harris (trans.). Indianapolis: Hackett Publishing.

Heidegger, M. (1979). *Nietzsche, Volume I: The will to power as art*, D.F. Krell (trans.). New York: Harper & Row.
Kant, I. (1998). *Groundwork for the metaphysics of morals*. M.J. Gregor (ed. and trans.). Cambridge: Cambridge University Press.
Malabou, C. (2005). *The future of Hegel: Plasticity, temporality and dialectic*, L. During (trans.). New York: Routledge.
Nancy, J.-L. (2002). *Hegel: The restlessness of the negative*, J. Smith and S. Miller (trans.). Minneapolis: University of Minnesota Press.
Nietzsche, F. (1990). *Twilight of the idols*, R.J. Hollingdale (trans.). New York: Penguin Putnam.
Nietzsche, F. (1993). *Birth of tragedy*, S. Whiteside (trans.). New York: Penguin Putnam.
Sallis, J. (1991). *Crossings: Nietzsche and the space of tragedy*. Chicago: The University of Chicago Press.
Wood, A. (1990). *Hegel's ethical thought*. Cambridge: Cambridge University Press.

The Posthumanist Challenge to a Partly Naturalized Virtue Ethics

Roberta M. Berry

Virtue ethics has enjoyed a revival beginning in the mid-twentieth century and now is counted among the major normative theories. The predominant approach in contemporary virtue ethics incorporates ethical naturalism, but does not aspire to a wholly naturalized, objective account.[1]

Challenges have been posed to the viability of a revived contemporary virtue ethics, including its justification of the virtues, action-guiding capacity, capacity to cope with dilemmas, apparent self-centeredness, apparent cultural relativity, and invocation of character traits in the face of evidence of "situationist" social psychology—and those engaged in the virtue ethics project have offered responses to these challenges.[2] In this chapter, I assume the potential viability of the project of virtue ethics in the face of challenges that have been lodged to date, and I focus on the potential viability of partly naturalized approaches in the face of the emerging challenge of posthumanism.

In Part I, I sketch two influential partly naturalized accounts. My treatment centers on the ways, somewhat different, in which they might be said to be naturalized. I conclude by noting why I refer to them as "partly naturalized" and by noting the appeal of these partly naturalized accounts.

Part II outlines an emerging challenge to these accounts: the potential development and use of genetic engineering or other technologies to create "posthumans"—beings derived from and similar to humans but radically different in their capacities, temperaments, or embodiments. I explore the ways in which the availability and use of these technologies might undermine the viability of the project of partly naturalized virtue ethics.

I consider a range of potential responses by a partly naturalized virtue ethics to the posthumanist challenge in Part III, concluding that none of them is wholly adequate to the challenge but one of them goes part way. And, in Part IV, I suggest that the development of a "virtue politics" is needed if a more adequate response is

R.M. Berry (✉)
School of Public Policy, Georgia Institute of Technology, Atlanta, GA, USA
e-mail: robertaberry@gatech.edu

M.J. Cherry (ed.), *The Normativity of the Natural*, Philosophical Studies in Contemporary Culture 16, DOI 10.1007/978-90-481-2301-8_11,
© Springer Science+Business Media B.V. 2009

to be possible, and I sketch what I call a "navigational" approach to policymaking as a component of that political theory.

1 Partly Naturalized Virtue Ethics

Two influential virtue ethics scholars, Christine Swanton and Rosalind Hursthouse, invoke naturalism in somewhat different ways in their partly naturalized accounts of virtue ethics, as explained below. In brief, neither claims a foundation for their accounts external to ethical beliefs. Instead, Swanton invokes a coherentist justification of the virtues in which a background theory of human nature contributes to a Rawlsian "wide reflective equilibrium." In Swanton's view, the virtues are states of "appropriate responsiveness" to the "demands of the world"; what is "appropriate" is partly shaped by the concept of flourishing. Hursthouse invokes a stronger neo-Aristotelian naturalism in which the virtues are those character traits that conduce to flourishing and are partly constitutive of it. Her concept of flourishing is not drawn from a wholly naturalized, scientific account of human nature; instead, it is formulated from within an ethical point of view that is grounded in a scientific account of humans as rational social animals (Swanton, 2003; Hursthouse, 1999, 2007).

Christine Swanton, in *Virtue Ethics: A Pluralistic View*, adopts Aristotle's conception of virtue as "a disposition in which both reason and emotion are well ordered" (2003, p. 8). The virtuous agent has "practical wisdom, right ends which are both expressed in and promoted by her actions, and correct affective states" (p. 8). Also from Aristotle, Swanton takes the idea that virtue is a state of appropriate responsiveness to what she calls "the demands of the world" and that the concept of flourishing partly shapes what is appropriate (p. 8).

Swanton, though, wants to move beyond what Aristotle thought adequate to normative theory: rendering coherent "the beliefs of the many or the wise." On her coherentist approach, she aims instead to achieve a wide reflective equilibrium among beliefs and a background theory about human nature (pp. 8–9). Aristotle's background theory of human nature, Swanton notes, is inadequate—"we do not, in Aristotle, understand virtues via an account of flourishing—we understand flourishing via an account of the virtues" (p. 9). Virtue ethics, in Swanton's view, needs to

> take seriously the incorporation of background theory into virtue ethics. Uncodifiability [the inability to reduce action guidance to a principle or set of principles to be applied by a moral agent] is not just founded in the intricacies of practical wisdom; it is sourced in the extreme difficulty of knowing how to distinguish virtue itself from (very closely allied) vice. ... Once we see the moral universe through the prism of the language of virtue and vice, we see that uncodifiability is not just due to the complexity of situations, it is due to the complexity of virtue and vice itself, and ultimately to the complexity (and opacity) of human nature (pp. 9–11).

Swanton explains that the background theory employed in her project is "not to be understood as formed from a value-free 'empirical' standpoint which renders it 'neutral' and thereby privileged. On the contrary, the psychology constituting such theory is itself deeply philosophical, and is actively informed by the thinking of philosophers such as Aristotle, Nietzsche, and Kierkegaard" (p. 13). But the background theory goes beyond the "insights and speculations of the philosopher-psychologists" to include the writings of "psychologists who emphasize the importance of personality and character" because these writings are "informed by a training and a clinical experience unavailable to the philosopher as such" (p. 13). As with virtue ethics, these writings presuppose that "personality and character are a central feature of human nature and human ethics, that human nature is to be understood in a holistic way, and that human beings should be understood via a psychology which recognizes their nature as beings who grow and mature" (p. 13). While Aristotle also recognizes this, he lacks an appreciation of what constitutes healthy human growth and "the vicissitudes which interfere with it of the kind we see in post-Nietzschean psychology" (p. 13).

Accordingly, Swanton applies what she calls a "Constraint on Virtue": "what counts as a virtue is constrained by an adequate theory of healthy human growth and development" (p. 15). But she rejects the stronger view of Rosalind Hursthouse, among others, that a necessary condition for a virtue is that it contribute to or partly constitute the flourishing of the agent (pp. 15, 19, 77–90): "On my own view, it is possible for a trait to be a virtue even where it has tendencies to be bad for the possessor in non-utopian universes such as our actual world" (p. 20).

Rosalind Hursthouse develops this stronger view in her influential *On Virtue Ethics* (1999), which builds on the work of a founder of contemporary virtue ethics, Philippa Foot.[3] On Hursthouse's account, the virtues conduce to human flourishing and possession of the virtues renders their possessor a good specimen of humankind. Hursthouse recites evidence for the claim that the virtues contribute to flourishing (1999, pp. 20–21, 163–191): the benefits of honesty, for example (p. 168); that we choose to teach our children to be virtuous; that we strive to be virtuous ourselves (pp. 20, 174–177). While this sort of evidence does not amount to "objective justification" for the claim that the virtues benefit us (pp. 163–165), from within an ethical perspective, we can know this to be true (pp. 189–191).

Hursthouse also recites evidence in support of her second claim, that the virtues are traits that make one "good qua human being" (1999, p. 20). She compares humans to other social animals and identifies a common set of purposes to: (1) survive, (2) procreate or otherwise help perpetuate the species, (3) pursue one's enjoyments, and (4) support one's social groups, which, in turn, support the pursuit of the first three purposes. Humans, distinctively, exercise reason in pursuit of these purposes and fulfill them in far more diverse ways than nonhuman animals—this precludes a deterministic, wholly naturalized virtue ethics, Hursthouse explains. An account of the virtues must respect these naturalized constraints but the account can be developed only from within an ethical perspective (1999, pp. 21, 229; see generally 195–238).

Hursthouse claims, further, that the virtues both benefit us and make us good specimens of our kind due to a harmoniousness in human nature (1999, pp. 247–265):

> It does not just happen to be the case that those character traits which benefit their human possessor, enabling her to live a satisfying and fulfilling life, coincide with those character traits which are the good-making characteristics of human beings. They benefit her in this way because of her nature as a human being, the sort of rational social animal that human beings are (p. 251).

Again, from within an ethical perspective that assumes this harmoniousness in human nature, we can recognize the possibility of this coherence and observe its operation in human life (pp. 259–265).

In a 2004 discussion of her project, Hursthouse focuses on the naturalism of her account and the Aristotelian claim that "considerations of human nature provide, in some sense, a rational foundation for ethics" (2004, p. 263). Can this claim succeed in light of Hursthouse's acknowledgement that a virtue ethics account must be built from within an ethical perspective?

Hursthouse draws on the work of John McDowell to address an apparent trilemma posed to the naturalist project (2004, p. 265): (1) the project supplies an objective grounding for virtue ethics only at the price of an unacceptable determinacy: what is natural is good; reproduction is natural for women; women who fail to reproduce are defective; (2) the project supplies an objective grounding that is unacceptably *in*determinate, offering a biological, ethological account of human nature that yields little in the way of ethical content; (3) the project supplies an account of human nature formulated from within an ethical outlook and, while it might yield a bounty of ethical content, it does so at the price of an unacceptable failure of objectivity; its ethical content was contained in the ethical outlook. Hursthouse considers whether her approach—which rejects (1) the strong determinacy of the natural as normative and (2) the indeterminacy of a descriptive ethological account—must then retreat into (3): merely reiterating the normative content of an ethical outlook embraced at the outset, surrendering any claim to objectivity grounded in human nature.

To address this, Hursthouse follows McDowell in rejecting the view that ethical conclusions—or scientific conclusions for that matter—are derived "top-down" from "a few independently established premises" (2004, p. 266). Rather, we proceed from within a conceptual scheme that "embraces our ethical outlook" in Neurathian fashion, "scrutinizing it, validating or changing it, bit by bit, plank by plank" (p. 266). This approach, Hursthouse argues, creates "a space . . . within which an enterprise of ethical naturalism could proceed, unhampered by a false scientism" (p. 266). If we keep in mind the Neurathian image of a ship constantly undergoing revision as it sails, "you can see the possibility of radical ethical reflexion, the critical scrutiny of one's ethical views which can be genuinely radical and not merely a reiteration of an acquired ethical outlook despite proceeding from within it" (p. 266). This is the project that Hursthouse aspires to pursue, supplying an account of human nature that proceeds within an ethical outlook but that can lay claim to objectivity.

A project of this sort, Hursthouse cautions, will not yield a set of moral rules or provide "crisp conclusions", about the rightness or wrongness, for example, of abortion. Instead, if it succeeds, it will supply reasons, anchored in human nature, to believe that certain traits constitute what we widely recognize to be the virtues

(2004, p. 267). From the premise that a trait constitutes a virtue, we cannot proceed directly to a conclusion about whether a particular course of conduct constitutes acting virtuously or not. Only after "a great deal of thought, casuistry and *phronesis* [practical judgment]" applied to the particulars of the case can we hope to decide this (pp. 267–268).

This naturalized approach, Hursthouse notes, can and should accommodate wide variability in the ways in which virtuous humans lead their lives. A partly naturalized virtue ethics does not prescribe that all must lead the same sorts of lives to be good human beings and to flourish, only that they live in accordance with the virtues (2004, p. 271).

This approach acknowledges that some of our ethical disagreements will be due to disagreement about certain facts—and that these disagreements are likely to persist. For example, theists and non-theists may agree about the four purposes shared by human and nonhuman social animals, but will disagree, due to their disagreement about certain metaphysical facts, about whether additional human purposes and corresponding virtues must be included in a partly naturalized virtue ethics account (2004, pp. 271–272). Finally, Hursthouse notes the reliance of her account on appeal to facts that are neither empirical facts, accessible by neutral observation or analysis, nor "moral facts," accessible only from within a particular ethical outlook. The ethical outlook from within which her account is built includes certain "ethical but non-evaluative beliefs about human nature and how human life goes" (p. 275), such as that crime does not pay or that money does not bring happiness or that you cannot fool most of the people most of the time (pp. 273–275). She notes the role of these beliefs in her account "both for their intrinsic interest and also as a detailed illustration of some of what might be called the limitations of the proposed naturalistic project" (p. 275).

I refer to these accounts as "partly" naturalized to call attention to the limits of the naturalism they invoke, as discussed above. This might be seen as weakness or as strength; the former for those who subscribe to a naturalistic determinacy—or who reject naturalism altogether; the latter for those who see promise in an account that can lay some claim to objectivity and universalism while acknowledging the pervasive and persistent diversity that characterizes the moral life across time and place.

2 From Transhumanist Enhancement to the Posthumanist Challenge

Potential revisions to human beings that might be accomplished beginning as soon as the early decades of the 21st century can be arrayed across spectrums. A spectrum might be demarcated by type of intervention: pharmacological, somatic-cell gene therapy, germ-line genetic engineering, brain-machine interface technologies, and so on. Or it might be demarcated by characterizations of interventions as "therapeutic" versus "enhancement"; the former aimed at remedying recognized diseases or disabilities, the latter at enhancing or augmenting "normal" human functioning.[4]

For purposes of this chapter, I will sketch a spectrum that runs from "transhumanist" to "posthumanist" interventions, with the demarcation blurry in the middle and indeterminate with respect to the proper characterizations of some interventions. I use this metric to serve the purpose of this discussion: to identify those interventions that could result in beings radically different in ways potentially significant to a partly naturalized account. These interventions might matter for the account of Swanton, for example, if they result in fundamental changes to the account of healthy growth and development included in her background theory. The interventions might matter for Hursthouse's account, for example, if they result in fundamental changes to the four shared human purposes she identifies or if they might disrupt the postulated harmoniousness that explains why virtues conduce to human flourishing and render their possessor good *qua* human being.

Nick Bostrom, Oxford philosopher and co-founder of a transhumanist association,[5] characterizes transhumanism as "an outgrowth of secular humanism and the Enlightenment" that views human nature as "improvable" (2005, 2003, p. 4). Humanism, on Bostrom's account, focuses on the significance of the individual and aims at improving the human condition through "rational thinking, freedom, tolerance, democracy, and concern for our fellow human beings" (2003, p. 4). While humanism envisions progress by application of reason to improve the world around us and to improve ourselves through educational and cultural development, transhumanism looks beyond this: what is our potential and how might we realize it by the application of reason to enhance the human organism? (p. 4). Bostrom claims that our current biological particulars do not capture "what is valuable about us"; instead, it is our "aspirations and ideals, our experiences, and the kinds of lives we lead" that are valuable (p. 4). Hence, progress will follow when more people are empowered to choose to "shape themselves, their lives, and the ways they relate to others, in accordance with their own deepest values" (p. 4).

Bostrom envisions the future fruits of current research:

> New kinds of cognitive tools will be built that combine artificial intelligence with [brain] interface technology. Molecular nanotechnology has the potential ... to give us control over the biochemical processes in our bodies, enabling us to eliminate disease and unwanted aging. Technologies such as brain-computer interfaces and neuropharmacology could amplify human intelligence, increase emotional well-being, improve our capacity for steady commitment to life projects or a loved one, and even multiply the range and richness of possible emotions (p. 4).

Research toward technologies that might accomplish some of these revisions is discussed and evaluated in a National Science Foundation sponsored report of a December 2001 workshop on the convergence of nanotechnology, biotechnology, information technology, and cognitive science ("NBIC") (World Technology Evaluation Center, 2002). The repot's editors note that participants in the workshop envisioned "important breakthroughs in NBIC-related areas in the next 10 to 20 years" and "significant applications" in the decades following. The NBIC Report describes some of these potential breakthroughs and applications:

> Sustaining human physical and mental abilities throughout the life span would be facilitated by progress in neuroscience and cellular biology at the nanoscale. An active and dignified

life could be possible far into a person's second century, due to the convergence of technologies. Gene therapy to cure early aging syndromes may become common, giving vastly improved longevity and quality of life to millions of people.

New communication paradigms (brain-to-brain, brain-machine-brain, group) could be realized in 10–20 years. Neuromorphic engineering may allow the transmission of thoughts and biosensor output from the human body to devices for signal processing. Wearable computers with power similar to that of the human brain will act as personal assistants or brokers, providing valuable information of every kind in forms optimized for the specific user (World Technology Evaluation Center, 2002, pp. 18–20, internal citations omitted).

With advances in the latter decades of the 21st century, the NBIC Report envisions far-reaching implications for society as well as individuals:

People may possess entirely new capabilities for relations with each other, with machines, and with the institutions of civilization.... A networked society of billions of human beings could be as complex compared to an individual human being as a human being is to a single nerve cell. From local groups of linked enhanced individuals to a global collective intelligence, key new capabilities would arise from relationships created with NBIC technologies. Such a system would have distributed information and control and new patterns of manufacturing, economic activity, and education. It could be structured to enhance individuals' creativity and independence. Far from unnatural, such a collective social system may be compared to a larger form of a biological organism....

Human culture and human physiology may undergo rapid evolution, intertwining like the twin strands of DNA, hopefully guided by analytic science as well as traditional wisdom. ... [T]he pace of change is accelerating, and scientific convergence may be a watershed in history to rank with the invention of agriculture and the Industrial Revolution (World Technology Evaluation Center, 2002, pp. 21–22).

Many transhumanists, on Bostrom's account, hope to live long enough to take the next step—to transform themselves into "posthumans", whose "basic capacities so radically exceed those of present humans as to be no longer unambiguously human" (Bostrom, 2003, pp. 5–6). Posthumans might enjoy "indefinite health-spans, much greater intellectual faculties than any current human being—and perhaps entirely new sensibilities or modalities—as well as the ability to control their own emotions" (Bostrom, 2005). Bostrom envisions that we might be capable of enhancing our children through genetic engineering by some point in the mid-twenty-first century. But, likely sooner than that, we will be capable of significant or radical self-revision by applications of nanomedicine, artificial intelligence, uploading of human intellect to a computer, or somatic-cell gene therapy (2003, p. 22). Posthumans might consist of "synthetic artificial intelligences", biological brains uploaded to computers, or beings resulting from a series of small but "cumulatively profound augmentations" to biological humans (2003, p. 5).

Bostrom notes, however, that we should guard against the creation of posthumans that threaten our existence or that deprive most of us of the benefits of posthuman revision:

[I]f and when superintelligence is created, it will be of paramount importance that it be endowed with human-friendly values. An imprudently or maliciously designed superintelligence, with goals amounting to indifference or hostility to human welfare, could cause our extinction. Another concern is that the first superintelligence, which may become very powerful because of its superior planning ability and because of the technologies it could

> swiftly develop, would be built to serve only a single person or a small group (such as its programmers or the corporation that commissioned it). While this scenario may not entail the extinction of literally all intelligent life, it nevertheless constitutes an existential risk because the future that would result would be one in which a great part of humanity's potential had been permanently destroyed and in which at most a tiny fraction of all humans would get to enjoy the benefits of posthumanity (2003, p. 24).

What is the likelihood that we might engage in these sorts of transhumanist or posthumanist revisions? As evidence of the possibility, we can point, first, to technologies currently in use or in prospect that fall short of transhumanist revision, but that evidence our capacity to invent and willingness to make use of techniques for revision.

For example, currently, by employing screening procedures in conjunction with in vitro fertilization (IVF), parents can select for or against certain features in their future children before birth, such as sex, deafness, or vulnerability to genetically influenced conditions (President's Council on Bioethics, 2003, pp. 28–44; Berry, 2006a, b, 2007). Also, parents can employ psychotropic drugs after their children are born to revise behavioral features of their children, whether as treatment for attention-deficit/hyperactivity disorder (ADHD), depression, or obsessive-compulsive disorder or, on the assessment of some commentators, as revisions in behaviors associated with these conditions but not properly characterized as amounting to these conditions (President's Council on Bioethics, 2003, pp. 28–44; Berry, 2006a, b, 2007).

Currently, steroids can be self-administered for muscle enhancement and other agents can be taken to improve attention and stamina. Additional, more sophisticated biotechnological enhancement agents are in prospect (President's Council on Bioethics, 2003, pp. 101–131; Berry, 2006a). Current drugs also can blunt traumatic memories and revise mood, forestalling posttraumatic stress disorder or treating depression—or relieving us of bad memories or making us feel better than well—with research on additional agents underway (President's Council on Bioethics, 2003, pp. 205–225; Berry 2006a). And research on anti-aging agents that might significantly extend the current maximum human lifespan of about 122 years—with the prospect of long, healthy lives extending across multiple generations—is ongoing (President's Council on Bioethics, 2003, pp. 159–181; Berry, 2006a). Ethical and policy discussion with respect to the use or potential use of these technologies is well underway (President's Council on Bioethics, 2003, pp. 131–156, 181–202, 225–270; Rasko, 2006).

Secondly, we can point to commentary envisioning interventions that approach or fall squarely within the bounds of transhumanist intervention. Commentators envision that parents might one day elect genetic engineering in conjunction with IVF to revise features of their future children, such as intelligence (Stock, 2002, pp. 97–123, 179), or to engineer for novel features, such as infrared vision (Silver, 1998, pp. 243–247, 277–293; Stock, 2002; President's Council on Bioethics, 2003, pp. 37–40; Berry 2006a, b). Genetic engineering might also enable parents to revise behavioral features of their children before birth, to reduce or increase assertiveness or aggression, for example (Glover, 1984). Ethical and policy debate regarding the prospect of these genetic engineering interventions also is well underway (see, e.g., Berry, 2007; Rasko, O'Sullivan and Ankeny, 2006; Silver, 1998; Stock, 2002). Some doubt the imminence of the techniques, while nonetheless expressing concern about the prospect (President's Council on Bioethics, 2003, pp. 37–40; Kass, 2002, p. 121). Others view the development and use as inevitable and turn their attention to how we should view and address the results (Baylis and Robert, 2006; Juengst, 2006, p. 136).

Regarding the prospect of posthumanist interventions, views are mixed both regarding the means and the motivation to engage in them. Views range from doubt about the imminence but concern about the prospect (Fukuyama, 2002, pp. 72–83) to cautious speculation (Baylis and Robert, 2006) to near-certainty that the capacity to engage in posthumanist revision is close at hand and enthusiasm about the prospect:

> The singularity is a metaphorical social event horizon in which accelerating technological trends so change society that it is impossible to forecast what the world will really be like. [Ray] Kurzeil believes that humanity will accelerate itself to utopia (immortality, ubiquitous [artificial intelligence], nanotech abundance) in the next 20 to 30 years. For example, he noted that average life expectancy increases by about 3 months every year. Kurzeil then claimed that longevity trends are accelerating so fast that the life expectancy will increase more than one year for each year that passes in about 15 years. In other words, if you can hang on another 15 years, your life expectancy could be indefinitely long. He projects that by 2030, [artificial intelligence] will be ubiquitous, and most humans will be physically melded to information and other technologies. Kurzeil argued that we must reject the fundamentalist desire to define humanity by its limitations. "We are the species that goes beyond our limitations," he declared (Bailey, 2007; see also, Bostrom, 2003, 2005).

I assume in this chapter that the development and use of transhumanist interventions are quite likely and I argue that, as the nature of the envisioned revisions approaches posthumanist, they will be potentially challenging to partly naturalized virtue ethics accounts. I assume that the development and use of posthumanist interventions, involving the most profound revisions to human capacities, temperament, and embodiment, are within the realm of possibility—and I argue that these interventions assuredly would be challenging to these accounts.

What are the challenges that might be posed to partly naturalized virtue ethics accounts by the potential development and use of near- and clearly posthumanist technologies? Could posthumans be assimilated to a partly naturalized virtue ethics account that aspires to some measure of objectivity as well as universality—that now would extend not only across time and place, but species or kind? There are reasons to doubt this could be accomplished.

As opaque as human nature may be to us at times and in some regards, posthuman nature or natures are far more difficult for us to fathom, at least from a distance, and likely close up: potentially immortal, nearly invulnerable, vastly intelligent synthetic artificial intelligences, uploaded biological brains, or augmented biological humans, perhaps with reworked temperaments and networked communication. We surely could conclude that their developmental portrait, whatever else it might be, would be different from the account of healthy human growth and development that Swanton asserts should shape our conception of the virtues.

On Hursthouse's account, we might wonder whether the purposes she identifies for humans, as rational social beings, would necessarily hold for some or all posthumans. Posthumans might remain concerned with their own survival, if individual identity were still characteristic of their lives. But if immortality were assured, barring accident or assault, this purpose might take on a rather different hue. Whether they would show an interest in something like procreation or perpetuation of their own species or of the set of hominid species that includes humans and some or all varieties of posthumans or instead would be interested in ensuring only their own individual immortality is difficult to know. We might envision them as concerned with their enjoyments much like us, but the nature of these enjoyments likely would extend far beyond the diversity that Hursthouse strives to accommodate or any of us currently can imagine. Their interest in supporting social groups would seem to turn crucially on their need for and interest in the support of these groups under circumstances of radically revised intellects, temperaments, embodiments, and, potentially, social relations. There is reason to doubt that the concept of flourishing would be the same for humans and for posthumans or that the same virtues would conduce to their flourishing or render them good specimens of their kinds.

Even if we were to assume a human-posthuman commonality across these four purposes, could we also assume a harmoniousness in posthuman nature that would ensure that the virtues conduce to the flourishing of posthumans and render them good members of our joint kinds? Or, could we design posthuman beings who would experience healthy posthuman growth and development that could be used to shape the virtues in ways consistent with Swanton's Constraint on Virtue? Perhaps these questions could be answered by the designers of posthumans, who might strive to engineer the biological or synthetic components of a harmoniousness of the sort that Hursthouse posits for humankind or of healthy growth and development on Swanton's account.

The engineering accomplishment would require the capacity to anticipate the consequences of design efforts in the creation of beings different from and at least as complex as ourselves. Perhaps we could succeed in this, especially after transhumanist enhancement of our intellect and given that we currently understand our engineered projects, such as bridges and airplanes, better than ourselves. But perhaps not, if we consider the current limits on our capacity to comprehend our own complexity.

Philosopher of biology Alexander Rosenberg notes that we are the product of a selective evolutionary process and, as such, are well adapted to understanding the demands of the world we inhabit, but, at the same time, not competent to comprehend the full complexity of our biological selves. We are more complex than we can comprehend due to selection on the basis of function or purpose rather than structural kind: a huge profusion of structures can and do enable us to perform those functions necessary for us to survive and reproduce (1995, pp. 143–155). Whether we, or our transhuman successors, would be competent to engage in the most complex engineering project imaginable—to render our posthuman nature harmonious or ensure our posthuman healthy growth and development—is, at best, questionable.

The posthumanist challenge calls into question the appeal of a partly naturalized virtue ethics in a posthumanist era. Could the promise of a partly naturalized virtue ethics—to claim some measure of objectivity and universalism while acknowledging the diversity that characterizes the moral life—be sustained?

3 Potential Responses to the Posthumanist Challenge

I consider here four possible responses to the posthumanist challenge. I note the shortcomings of each, while acknowledging the partial success of the last.

1. Deep ethical naturalism. On this response, we would understand the presuppositions of a partly naturalized virtue ethics, whether on the weak version of Swanton or the strong version of Hursthouse, to include a deep commitment to the normativity of the natural in the following sense: human nature must be preserved as a necessary foundation to ethics. Human nature is the foundation that permits some measure of objectivity as well as universality. While we can and should revise the world and reform ourselves in pursuit of a virtuous life, we should not revise the anchor that makes these possible. Accordingly, posthumanism should be rejected as a matter of policy—that is, we should not support or encourage and we should perhaps discourage or prohibit efforts to develop posthumanist technologies—and, if these technologies should nonetheless become available, virtuous moral agents should abstain.

   Perhaps, on Hursthouse's account, we might claim that this principle is part of an ethical outlook within which a project of partly naturalized virtue ethics is built. But it would be difficult to characterize a principle of preservation of human nature, as bequeathed to us by our evolutionary past, as simply an aspect of an ethical point of view. We might just as well appeal to a principle of preservation of human nature as crafted by God. In either case, the claim to objectivity and universality grounded in human nature is surrendered—we can be certain of disagreement about the normativity of the natural or the existence and role of God—and the debate extends well beyond disputed facts, natural and metaphysical, about human nature. Resort to an external principle of deep ethical naturalism is inconsistent with the appeal of a partly naturalized virtue ethics.
2. Prudential preservationism. Because the virtues are shaped by or conduce to human flourishing, on this response, we should not tamper with conditions, whether of our natural or social environment or ourselves, that would threaten to unsettle these relationships. We should oppose, as a policy matter, the development or use of posthumanist technologies and refrain as a matter of individual choice.

   This response essentially invokes a risk-benefit calculus under conditions of uncertainty. But we have engaged in myriad activities, from the establishment of schools that separate children from parents for substantial periods of time in their early years, to the invention and administration of psychotropic drugs, to the embrace of life-saving medical interventions—despite the warnings of eugenicists that this would undermine our flourishing by increasing the "genetic load." On what basis might we separate these and other interventions, each with attendant bundles of risks and benefits under conditions of uncertainty, from posthumanist interventions? As with these other interventions, we can only proceed from the virtues, guided by practical judgment, on a case-by-case basis, to evaluate the risks and benefits as best we might. A partly naturalized account of virtue ethics cannot recognize an in-principle prudential objection to interventions that are undertaken by individuals as a matter of choice because we believe they are likely to revise aspects of human nature that ground virtue ethics—without importing a *sub rosa* principle of deep ethical naturalism.

3. Pluralistic Naturalism. Perhaps a partly naturalized virtue ethics can embrace the possibility of a variety of posthuman natures, each the subject of a background theory or foundation for a partly naturalized posthuman account of virtue ethics. Posthuman virtue ethicists, as human virtue ethicists, would engage in justifying the virtues as shaped by posthuman flourishing or according to whether they conduce to posthuman flourishing. And virtuous posthuman moral agents would act from the relevant set of virtues.

   On this approach, we might have multiple versions of partly naturalized virtue ethics: PNVE-H involving virtues shaped by our understanding of human nature or determined by their conducing to and partly constituting human flourishing; PNVE-PH1 involving the same with respect to posthumans constituted of biological humans who have been genetically engineered in certain ways to increase their capacities and render them immortal; PNVE-PH2 involving the same with respect to posthumans genetically engineered in certain other ways; PNVE-PH3 involving the same with respect to posthumans constituted of mortal humans greatly augmented in certain ways by nanotechnologies; PNVE-PH4 involving the same with respect to posthumans consisting of biological brains uploaded to computers who are immortal; PNVE-PH5 involving the same with respect to posthumans consisting of artificial intelligences; and so on.

   "Pluralistic naturalism" might be likened to "serial pluralism," if we envision that humankind over evolutionary time and extending into the future has encompassed and will encompass moral agents with natures sufficiently different to require different virtue ethics accounts: PNVE-H1, PNVE-H2, and so on. Both serial and pluralistic naturalism presuppose an external principle that holds across all hominid species, including those of human design, and that ensures a relationship among the virtues and flourishing like that set forth in current-day accounts of partly naturalized virtue ethics. But pluralistic naturalism raises the prospect, as serial pluralism does not, of the viability of a partly naturalized virtue ethics that relies on this presupposition: could we refrain from judging harshly a posthuman who acted consistently in ways that would be vicious on a current-day partly naturalized virtue ethics account tailored to humans but virtuous on a posthuman account? Or would we judge the posthuman conduct as "vicious" nonetheless—or perhaps the posthuman nature that anchored the account as "vicious-in-kind"?

If we refrained from harsh judgment of the posthuman conduct or agent, in accordance with this external presupposition, we would be acknowledging a meta-objectivity and meta-universalism that privileges the relationship among virtue, flourishing, and human or posthuman nature. In so doing, however, we would be forced to surrender the objectivity and universalism that previously pertained to contemporaneous moral agents in this world, abandoning the appealing features of a partly naturalized virtue ethics. On the other hand, if we judged the posthuman conduct or posthuman agent harshly, we would be acknowledging a parochialism in our account that privileges human nature, once again invoking *sub rosa* a principle of deep ethical naturalism.

4. Virtuous Choice. On this approach, there is no theoretical response called for from a partly naturalized virtue ethics. There are only the responses of moral agents acting from the virtues when confronted with choices. In deciding whether to support the development of posthumanist technologies, or, if those technologies are available, whether to use them, an appropriate responsiveness to the demands of the world, on Swanton's account, or virtues that conduce to our flourishing, on Hursthouse's account, might well lead individuals to refrain from participation in the development of posthumanist technologies or their use in the particular circumstances of ordinary life. In the event, for example, of an epidemic or natural catastrophe that threatened our survival, their judgment might be different.

   The shaping of the virtues according to healthy human growth and development or the determination of the virtues according to whether they conduce to our flourishing would likely lead us to a preservationism in fact, acting from a cluster of virtues: love, caring, loyalty, and commitment to ourselves, our children, our communities; caution, prudence, and humility regarding the complexity of the contemplated interventions, the risks, and their irreversibility; and the exercise of practical judgment about the *de minimis* or negative contribution of augmented powers or virtual existence, for example, to a life of flourishing.

   This response goes part way toward answering the posthumanist challenge in a way that does not undermine the appeal of a partly naturalized virtue ethics, but instead underscores it. But it goes only part way because it cannot adequately address the significance—to our capacity, as moral agents, to choose virtuously and live lives of flourishing—of choices at the level of the political community.

4 The Need for a Political Theory of the Virtues: A Navigational Approach

Accounts of virtue ethics recognize that our natural and social environments can support or diminish our capacity to act virtuously and flourish (Swanton, 2003,

p. 14; Hursthouse, 1997, p. 232). If, as I conclude, the best part-way response of a partly naturalized virtue ethics to the posthumanist challenge consists in virtuous choice by moral agents in response to the particulars presented by options to employ posthumanist interventions, how might the political community best ensure the opportunity to choose virtuously? Political communities, for example, might choose, by their policies, to promote or encourage, prohibit or discourage, or regulate in certain respects the development and use of these technologies. They might also devise policies regulating political rights and obligations among humans and posthumans. The challenge of posthumanism underscores the importance of developing a political philosophy of virtue ethics that can support virtuous individual choice in a future world in which the possibility of posthumanist revision might well become real.[6]

I aim here to sketch a conception of virtue politics on rough analogy to virtue ethics: how might we justify certain political—and policymaking—virtues that are shaped by or conduce to the flourishing of the political community? Political flourishing is understood, on a virtue politics account, to include, at a minimum, the opportunity for virtuous moral choice within pluralistic political communities. We might consider a range of contenders for a political philosophy that would satisfy this requirement.

Utilitarian political philosophy, generally, aims to maximize overall welfare—whether consisting in pleasure or happiness or satisfaction of preferences on various utilitarian approaches—after giving impartial consideration to each individual's welfare. Utilitarianism has a commonsensical appeal as an approach to accommodating the diverse welfare interests of all members of pluralistic political communities. Perhaps, then, a utilitarian account can ensure the flourishing of the political community, allowing the opportunity for virtuous individual choice regarding posthumanist technologies.

Political philosopher Robert Goodin notes that public utilitarianism, historically, has been viewed chiefly as a guide to policymaking rather than individual decision-making: "Jeremy Bentham was famous in his own day primarily as a reformer of legal systems; James Mill as an essayist on government; John Stuart Mill as an essayist, social reformer and parliamentarian; John Austin as a jurisprude" (Goodin, 1995, pp. 11–12, citation omitted). As applied to individual moral agents, utilitarianism is not very appealing, Goodin writes:

> No one wants to run one's life like Gradgrind, the Dickensian parody of a good utilitarian. Furthermore no one can. The calculative load imposed by utilitarian maximization would absorb all one's time and attention, leaving none for actually acting on the conclusions of the calculations. In personal life, most dramatically, there simply has to be more scope for considerations of uncalculating affection, standing rules of conduct and qualities of character (1995, p. 7, citation omitted).

But, Goodin observes, "Using the felicific calculus for micro-level purposes of guiding individuals' choices of personal conduct is altogether different from using it for macro-level purposes of guiding public officials' choices of general social policy" (p. 61). At the level of policymaking, the weaknesses of utilitarianism at the individual level might be translated into strengths. Acting from impartial benevolence, policymakers can dispassionately gather and analyze aggregate data about the welfare interests of members of the political community to formulate and apply welfare-maximizing policies for the benefit of all (Goodin, pp. 62–65).

But would public utilitarianism preserve the opportunity for virtuous choice by individuals? A utilitarian account reduces flourishing to welfare, measured as happiness or pleasure or the satisfaction of preferences. So, for example, the utilitarian calculus might count the experience of pleasure or happiness or the satisfaction of preferences of uploaded biological brains experiencing virtual environments. But the calculus would have no place for the "data" of virtuous choice to refrain from posthumanist revision—acting from loyalty, love, caring, loyalty, commitment, caution, prudence, or humility. Hence the calculus would likely guide policymakers to expand opportunities for the posthumanist experience of virtual reality rather than to ensure opportunities for virtuous choice to refrain. Utilitarian philosophers might appeal to principles beyond the utilitarian calculus, for example, to exclude consideration of the welfare experienced in virtual environments or to guarantee a "right" to virtuous choice. But, if these are to amount to more than ad hoc efforts to avoid unappealing results, they would require separate justification.

Perhaps, then, we should consider theoretical approaches that invoke policymaking from principles other than impartial maximization of the welfare consequences of policy choices. John Rawls, in his 1971 *A Theory of Justice*, imagines representative individuals in the "original position" behind a "veil of ignorance" regarding their own assets and ends engaging as equals to choose the basic principles of political institutions. He conceives of this choice situation as a procedure for implementing a Kantian conception of self-legislated, universalizable principles (Rawls, 1971, pp. 251–257). In his revised approach set forth in *Political Liberalism*, Rawls jettisons the Kantian underpinnings of *A Theory of Justice*, acknowledging that the moral commitments of pluralistic political communities will be diverse and will preclude agreement on a comprehensive Kantian moral theory (1993, pp. xviii–xix). He proposes instead a "political conception" in which the principles are drawn from our common political culture and consist in an overlapping consensus supported from within diverse worldviews (pp. xxi–xxxii). These principles, in priority order, guarantee equal basic rights and liberties, and allow social and economic inequalities only if equality of opportunity is assured and only to the extent that these inequalities benefit the least advantaged (pp. 5–6).

Ronald Dworkin proposes a stronger redistributive principle embedded in a comprehensive moral theory rather than in a political conception of justice (2000, pp. 4–5). Dworkin's two principles assert that the success of each individual is of objectively equal importance and that each has ultimate responsibility for the success of her own life (p. 5). It follows that policymaking should ensure that individuals' choices, rather than their economic circumstances or race or gender, for example, should determine their success (p. 6).

As these examples illustrate, the justifications for deontological accounts vary; how might we choose in a nonarbitrary way among them? We might expect that their implications for accommodation of virtuous choice by individuals will also vary, but how might we determine the comparative implications, for example, of a

Rawlsian or a Dworkian political equality for ensuring the opportunity for virtuous choice?

Regarding choosing among competing formulations, Alasdair MacIntyre suggests that the basis for choosing across the entire range of political theories is, in fact, contested. He claims that rival accounts of justice—Humian, Kantian, utilitarian, Thomist, and twentieth-century liberal—are accompanied by rival accounts of rationality for choosing among them (1988, p. 2). On his Thomist account, rival accounts are mistaken in assuming that a public philosophy can be derived from a set of first principles rationally available to all: "It is a Cartesian error ... to suppose that first by an initial act of apprehension we can comprehend the full meaning of the premises of a deductive system and then only secondly proceed to enquire what follows from them" (p. 174). In MacIntyre's view, the choice proceeds differently:

> In fact it is only insofar as we understand what follows from those premises that we understand the premises themselves. If and as we begin from the premises, our initial apprehension will characteristically be partial and incomplete, increasing as we understand what it is that these premises do and do not entail (pp. 174–175).

On MacIntyre's account, we recognize traditions to which we have the greatest affinity and in these traditions we see linked conceptions of justice and rationality that are associated with a conception of the good life (p. 389). These competing traditions then can be evaluated over time according to their relative capacity to cope with challenges posed to the political community (p. 398).

And H. Tristram Engelhardt, Jr. challenges the aspirations of all contemporary political theoretical approaches, and proposes a different principle for ensuring the opportunity for choice congruent with one's commitments. He observes that pluralistic political communities are constituted of "moral strangers" who do not share

> (1) common moral premises, rules of evidence, and rules of inference so that their moral controversies can be settled by sound rational argument, or (2) a common understanding of who is in moral authority, so that their moral controversies can be settled by a definitive ruling or process (2000, p. 46, n 6).

On Engelhardt's account, the only defensible foundation for a secular bioethics among moral strangers must be "grounded in content-less procedures for deriving authority from the permission of individuals, the only moral authority possible for collaboration among moral strangers" (p. 3). A society based on a "libertarian cosmopolitan moral understanding" can encompass "divergent moral visions, moral communities, and fragments of moral communities," permitting all "who are willing to live peaceably with others" to act on their deepest commitments:

> The libertarian cosmopolitan moral understanding provides a *modus vivendi* with moral force but without content for a world marked by a plurality of moral visions. In this sense it is cosmopolitan. Because it is a procedural morality, it does not offer content for a content-full view of the significance of reproduction, birth, suffering, dying, and death, nor does it have content to support non-procedurally directed understandings of character and virtues. ... The libertarian ethic is a moral point of view that lacks a vision of the good moral life (p. 42).

A virtue politics might draw on the forgoing to aim at a conception of political flourishing that, like Engelhardt's account, acknowledges the fact of pluralism and its challenge to any content-full conception of political flourishing. In a political community in which we can be certain that moral agents will disagree about whether it would be right to engage in posthumanist revision, a libertarian understanding can at least ensure that the option to engage in virtuous choice is preserved.

If we contemplate the particulars of the posthumanist challenge, preserving the opportunity for virtuous choice likely will require preventive policymaking measures as well. For example, the exercise of some posthumanist choices might yield beings who threaten the opportunity for continuing choice—or worse. Those who advocate posthumanist choices acknowledge this possibility and urge that we guard against it.

Preserving the opportunity for choice might also call for policies that preserve a social context in which the fruits of these choices can be realized. If the infrastructure that supports human procreation and parenting, realization of enjoyments, and support for and from communities withers because, for example, social resources are directed exclusively to supporting the infrastructure of posthuman enjoyment, meaningful opportunities for humans to flourish also would wither.

A virtue politics adequate to the posthumanist challenge, then, must be competent to address the problems of preserving meaningful virtuous choice in pluralistic political communities. Its political virtues will include commitments to the liberty, equality, and flourishing of all members of the political community. This will entail policies that preserve meaningful choice for all. Devising these policies will require bringing practical collective wisdom to bear in the concrete context of the challenges posed. This policymaking process, what I call a "navigational approach," will consist in an analogue at the social level of individual virtuous decision-making guided by practical judgment.

On a navigational approach, the policymaking process will require gaining an understanding of and coping with the novel, complex problems presented by the posthumanist challenge under conditions of diverse commitments in a pluralistic political community. I analogize the task to that of common law decision-making, in which judges, as policymakers, must "take the measure of novel policy problems ... and arrive at resolutions guided by our common values" (Berry, 2007, p. 176):

> Advocates draw on the facts, reasoning, and holdings of past cases, arguing that the principles applied in those cases should or should not apply to the current case because they are relevantly similar or dissimilar. Advocates also point to an anticipated future in which the principles they advocate for the resolution of this case might be applied, inviting judges to envision the soundness of the principles on the basis of their future application to similar facts. Judges immerse themselves in the facts and circumstances, take in these arguments appealing to common values, and devise resolutions sufficient unto the day. In the written opinions of courts, we see examples of accretions to our store of practical knowledge—solutions employing principles no broader than required to resolve the disputes at hand and subject to revision over time as they are tested by future disputes arising from subsequent experience (Berry, 2007, p. 177).

In a similar way, a navigational approach to policymaking will ask us to apply great effort to understanding and great modesty in resolving the problems presented. These "policymaking virtues" will aim at incremental resolutions, with the expectation that the policymaking process will be ongoing and iterative as we observe the results of our efforts (p. 178).

Acknowledgments I am indebted to Lara Denis for her comments on an early version of this work.

Notes

1. See Stohr (2006, pp. 23–26) regarding the trend to naturalism in contemporary virtue ethics. Naturalized virtue ethics differs from natural law theory, which holds, generally, that nature has certain ends that are knowable and that are good and normative. See also Iltis (2004, p. 115). Stohr notes that virtue ethics and natural law theories are usually treated as distinct (2006, p. 23). Some of the challenges to a partly naturalized virtue ethics discussed in this chapter, however, are similar to those posed to natural law accounts. See, e.g., Cherry (2004a, b) and Iltis (2004). Other, non-naturalized influential contemporary accounts of virtue ethics include that of Michael Slote (2001).
2. For responses to these and other objections, see Solomon (1997); Hursthouse (1999, pp. 17–18, 25–42; 2004, 226–234); Swanton (2003, pp. 6, 273–291). See also Hursthouse (2007).
3. For Foot's recent development of the project see her *Natural Goodness* (2001).
4. See discussions in President's Council on Bioethics (2003); World Technology Evaluation Center (2002); Berry (2007).
5. Humanity+ (formerly the World Transhumanist Association). See http://transhumanism.org/index.php/WTA/about/ (accessed December 12, 2008).
6. On the need for a virtue politics, see Hursthouse (2007); see also the discussion in Berry (2007, 122–180).

Bibliography

Alasdair M. (1988). *Whose justice? Which rationality?* Notre Dame: University of Notre Dame Press.

Bailey, R. (2007) Would you give up your immortality to ensure the success of a posthuman world? Answering hard questions at the World Transhumanist Conference. *Reasononline*. July 27. [Online] Available: http://reason.com/news/show/121638.html (accessed October 22, 2007).

Baylis, F. and Robert J.S. (2006). Radical rupture: exploring biologic sequelae of volitional inheritable genetic modification, in J. E. J. Rasko, G. M. O'Sullivan and R. A. Ankeny (eds.), *The ethics of inheritable genetic modification: A dividing line?* (pp. 131–148). Cambridge: Cambridge University Press.

Berry, R. M. (2006a). 'Beyond therapy' beyond the beltway: an opening argument for a public debate on enhancement biotechnologies. *HealthCare EthicsCommittee Forum* 18 (2): 131–155.

Berry R. M. (2006b). Can bioethics speak to politics about the prospect of inheritable genetic modification? If so, what might it say? J. E. J. Rasko, G. M. O'Sullivan and R.A. Ankeny (eds.), *The ethics of inheritable genetic modification: A dividing line?* (pp. 243–277). Cambridge: Cambridge University Press.

Berry, R. M. (2007). *The ethics of genetic engineering*. New York: Routledge.

Bostrom, N. (2003). The transhumanist FAQ: a general introduction, version 2.1. World Transhumanist Association. [On-line] Available: www.transhumanism.org/ resources/ FAQv21.pdf (accessed November 25, 2008).

Bostrom, N. (2005). In defence of posthuman dignity. *Bioethics* 19 (3): 202–214.

Cherry, M. J. (2004a). Natural law and moral pluralism, in M. J. Cherry (ed.), *Natural law and the possibility of a global ethics* (pp. 17–38). Dordrecht: Kluwer Academic Publishers.

Cherry, M. J. (ed.) (2004b). *Natural law and the possibility of a global ethics*. Dordrecht: Kluwer Academic Publishers.

Crisp, R. and Slote, M. (eds.) (1997). *Virtue ethics*. Oxford: Oxford University Press.

Dworkin, R. (2000). *Sovereign virtue: The theory and practice of equality*. Cambridge: Harvard University Press.

Engelhardt, H. T., Jr. (2000). *The foundations of Christian bioethics*. Lisse: Swets & Zeitlinger.

Foot, P. (2001). *Natural goodness*. Oxford: Clarendon Press.

Fukuyama, F. (2002). *Our posthuman future: Consequences of the biotechnology revolution*. New York: Farrar, Straus and Giroux.

Glover, J. (1984). *What sort of people should there be? Genetic engineering, brain control and their impact on our future world*. Harmondsworth: Penguin.

Goodin, R. E. (1995). *Utilitarianism as a public philosophy*. Cambridge: Cambridge University Press.

Hursthouse, R. (1997). Virtue theory and abortion, in R. Crisp and M. Slote (eds.), *Virtue ethics* (pp. 217–238). Oxford: Oxford University Press. First published in *Philosophy and Public Affairs* 20 (1991): 223–46. Also reprinted in D. Statman (ed.) (1997). *Virtue ethics: A critical reader* (pp. 227–244). Washington, D.C.: Georgetown University Press.

Hursthouse, R. (1999). *On virtue ethics*. Oxford: Oxford University Press.

Hursthouse, R. (2004). On the grounding of the virtues in human nature, in J. Szaif and M. Lutz-Bachmann (eds.), *What is good for a human being? Human nature and values* (pp. 263–275). New York: Walter de Gruyter.

Hursthouse, R. (2007). Virtue ethics, *Stanford encyclopedia of philosophy*. [On-line] Available: http://plato.stanford.edu/entries/ethics-virtue (accessed November 25, 2008).

Iltis, A. (2004). An assessment of the requirements of the study of natural law, in M. J. Cherry (ed.), *Natural law and the possibility of a global ethics* (pp. 115–122). Dordrecht: Kluwer Academic Publishers.

Juengst, E. T. (2006). 'Altering' the human species? Misplaced essentialism in science policy, J. E. J. Rasko, G. M. O'Sullivan and R. A. Ankeny (eds.), *The ethics of inheritable genetic modification: A dividing line?* (pp. 243–277). Cambridge: Cambridge University Press.

Kass, L. R. (2002). *Life, liberty and the defense of dignity: The challenge for bioethics*. San Francisco: Encounter Books.

President's Council on Bioethics (2003). *Beyond therapy: Biotechnology and the pursuit of happiness*. Washington, D.C. [On-line] Available: www.bioethics.gov/ reports/ beyondtherapy/ beyond_therapy_final_webcorrected.pdf (accessed November 25, 2008).

Rasko, J. E. J., O'Sullivan, G. M. and Ankeny, R. A. (2006). *The ethics of inheritable genetic modification: A dividing line?* Cambridge: Cambridge University Press.

Rawls, J. (1971). *A theory of justice*. Cambridge: Harvard University Press.

Rawls, J. (1993). *Political liberalism*. New York: Columbia University Press.

Rosenberg, A. (1995). *Philosophy of social science*, second edition. Boulder: Westview Press.

Silver, L. M. (1998). *Remaking eden: How genetic engineering and cloning will transform the American family*. New York: Avon Books.

Slote, M. (2001). *Morals from motives*. New York: Oxford University Press.

Solomon, D. (1997). Internal objections to virtue ethics, in D. Statman (ed.), *Virtue ethics: A critical reader* (pp. 165–179). Washington, D.C.: Georgetown University Press.

Statman, D. (ed.) (1997). *Virtue ethics: A critical reader*. Washington, D.C.: Georgetown University Press.

Stock, G. (2002). *Redesigning humans: Our inevitable genetic future*. Boston: Houghton Mifflin.

Stohr, K. (2006). Contemporary virtue ethics. *Philosophy Compass* 1 (1): 22–27.
Swanton, C. (2003). *Virtue ethics: A pluralistic view*. Oxford: Oxford University Press.
World Technology Evaluation Center (2002). Converging technologies for improving human performance: Nanotechnology, biotechnology, information technology and cognitive science." NSF/DOC-sponsored report, M. C. Roco and W. S. Bainbridge (eds.). [On-line] Available: http://www.wtec.org/ConvergingTechnologies/Report/NBIC_report.pdf (accessed November 25, 2008).

Part IV
The Challenge of Deriving an Ought from an Is

Can Moral Norms Be Derived from Nature? The Incompatibility of Natural Scientific Investigation and Moral Norm Generation

Ian Nyberg

1 Can We Derive Moral Norms from Nature?

Can moral normative principles or statements of moral normative value be derived from nature? One way to characterize this project is as a look into whether or not the naturalistic fallacy is in fact a fallacy. Some argue that nature is our best resource for acquiring moral norms. There are at least two different approaches to deriving morally normative content from nature. The first is the general use of nature to establish the fundamental principles of naturalistic normative theories. These theories posit that the good exists in nature, and that it is a discoverable, natural entity, perhaps, such as beauty or pleasure, or some other natural entity. This first approach lends itself to straightforward teleological theories where moral actions maximize the good. The second approach uses nature to find specific examples of moral norms; once these examples are found, moral actions are the ones that imitate or somehow cohere well with them.

Both of these approaches look for moral norms in nature using the same investigative methods used by the natural sciences. A possible third approach to ethical naturalism tries to avoid the uncertainty of the material world while also avoiding the uncertainty of the supernatural and intuitionist ethical approaches. Joseph Boyle's (1999) concept of natural law is an example of such an approach.

> [T]he notion of natural law: its core idea, as a thesis about morality, law and other forms of social authority is that some action-guiding thoughts and statements, that is, some precepts or practical principles, are natural in the sense that they are not dependent for their validity on human decision, authority or convention. Because of the independence of these factors, natural precepts and principles must be generally accessible to human reason; the critical reflection that is not dependent upon but potentially critical of any particular social enactment or practice is the work of common human reason. I will take this immediate implication concerning the accessibility of moral truth to human beings generally to be part of the core idea of natural law (Boyle, 1999, p. 2).

I am sympathetic to the desire for ethical realism, so I like Boyle's claim that there are ethical truths that exist independently of human judgment. Perhaps if there were

I. Nyberg (✉)
Department of Philosophy, University of Texas, Austin, TX, USA
e-mail: inyberg@mail.utexas.edu

M.J. Cherry (ed.), *The Normativity of the Natural*, Philosophical Studies in Contemporary Culture 16, DOI 10.1007/978-90-481-2301-8_12,
© Springer Science+Business Media B.V. 2009

some way to justify claims about the specific universal ethical truths Boyle speaks of, there would be a way to prove they exist. As it stands, there does not seem to be evidence, beyond our hope, for the existence of such truths. Likewise, there is no evidence that such truths do not exist. There does, however, seem to be substantial evidence against the claim that such ethical truths are universally accessible. The history of ethics offers a fair bit of evidence that even people trying to think carefully about ethics disagree about the fundamental ethical truths. Universal accessibility requires being able to show that the principles one is accessing are universally accessible. Without this condition, the claim of universal accessibility is empty, and functionally the same as intuitionist approaches. The focus of this article will be on naturalistic ethical theories that rely on the epistemology of the natural sciences.

Moral naturalism does not meet certain conditions desirable for an ethical theory because natural science is not suited to finding moral norms. One such condition is realism. In moral naturalistic theories, generally, ethical facts are the kinds of natural facts generated by the natural sciences (Lenman, 2006). Hence, the realism of moral naturalism would be scientific realism. It will be argued that scientific realism is untenable.

There seem to be only three possibilities for moral ontological space: (1) natural, where moral facts are natural facts, actual facts about the world that can be discovered with empirical investigation akin to the epistemic methods of the natural sciences; (2) non-natural, where moral facts are *sui generis*, for example in G. E. Moore's intuitionism; or (3) supernatural, for example, where moral facts are theological facts. One need not be a moral naturalist to hold that the fundamental good is pleasure, or beauty. Naturalism is just one way that philosophers have attempted to justify these concepts as fundamental to their respective ethical theories.

Because our scientific methods do not lend themselves to finding moral judgments, it will be argued that whether any particular judgment is in line with or against nature is ultimately independent of that judgment. In every case, we first have to judge what is good and what is bad in order to determine if the natural option is also the better option. This paper will argue that nature as such is neither a limit that bounds our ethical decisions, nor a challenge which our ethics must overcome.

1.1 Can We Discover the Good?

At first blush there is pragmatic appeal to turning to nature to ground an ethical theory. For one thing, ethical facts found in nature would seem to be more objective than those found if one of the other two possible moral ontologies turned out to be correct. For instance, intuitionist ethical theories, such as the theory G. E. Moore puts forward in his *Principia Ethica,* seem to necessitate private epistemologies. There is a substantial problem in that there is no non-question begging way to adjudicate between conflicting intuitions. Theological and other supernatural approaches, even when they appeal to the universality of human reason, often require preconditions such as faith or the giving of authority to sacred texts or religious committees, upon which our reason is meant to work. This is typified by St. Anselm's expression, "faith seeking understanding." Though reason is universal, particular religious commitments are not universal, which renders these ethical approaches, if not private, at least exclusive. It might be the case that one religion is in fact right about what it claims are the absolute ethical truths, and that all the others are wrong. So far, the position of virtually every religious group is that it is the one in the right. Likewise, it might be the case that one person's ethical intuitions are right and that everyone else's are wrong. Unfortunately, these positions are not helpful to everyone else until a way is demonstrated to determine which person or group is the right one. Disagreements between thoughtful, rational, faithful people about what the Deity is telling us about ethical truths are as difficult to adjudicate as disagreements between conflicting intuitions. The result is that supernatural approaches are functionally, epistemically equivalent to intuitionist approaches. These problems may make naturalism seem to be preferable for ethical theory building.

Consider the epistemology required for naturalism. What seems to be appealing about the moral naturalism project is at the same time its downfall; namely, that it utilizes the same empirical methods used to investigate nature within the physical and biological sciences. We will find that the epistemology sufficient for scientific theories is insufficient to give us satisfactory moral norms.

Naturalistic ethical theories are either in the realism camp or in some version of anti-realism, such as instrumentalism. Naturalistic ethical theories would be most applicable and universal, and therefore more desirable, if they were realist theories. Since the epistemology of naturalism is that of natural science, the type of realism in question is scientific realism. After all, we would want the ethical facts we found in nature to be natural facts and ethical truths. When we practice science, we think we are learning about the world in a certain sense. For instance, it does seem to be the case that we are better and better at manipulating the world and predicting certain phenomena. However, it is a mistake to think that contemporary science establishes scientific Truth (with a capital T). This may have been the intention historically, for instance in ancient and scholastic science. However, contemporary science recognizes that the predominant tool for our understanding of the natural world is induction. As has been well established since Hume, inductive inferences cannot be proven with the kind of epistemic certainty enjoyed by mathematics and mathematical logic. The effect this has on our understanding of the natural world is that empirical claims about the world are not necessarily true; they cannot be demonstrated absolutely. The best we can do is address how likely they are given how likely we think all of our other theories are.

This approach has proven very useful. So useful, in fact, that many epistemologists try to appeal to the usefulness of theories as evidence of their veracity. However, being careful requires that we keep in mind that there is no way to judge a theory's truth value other than its fit to the available data. Since we constantly gather new data, our theories are constantly susceptible to revision and refutation. The result is that scientific realism is unjustified. Popper seems to go so far as to suggest that we can only be certain that progress is being made when theories are disproved.[1]

If we applied this methodology for scientific demonstration and explanation, any ethical principles we derived would likewise be incomplete and inaccurate. It would be pretty unsatisfying if the only time we could be certain we were making ethical progress was when we threw out another theory.

The jump from having a theory that makes predictions with a certain degree of accuracy to saying that the theory also expresses some underlying deep truth about reality is unjustified. It is in principle impossible to prove that the phenomena or our observations of the phenomena will not change in the future. The best we can do is to estimate a probability that conditions will remain the same, making contemporary theories our best guess until new information comes along. The guarantee that new undermining information will come along is as good as the guarantee that no new undermining information will come along—nonexistent. Therefore, the realist belief that any given scientific theory is True cannot be knowledge (in any usual sense of knowledge, carefully speaking, where scientific knowledge is justified true belief, warranted true belief, or any standard model of knowledge pre or post Gettier (1963)).

There would be something more aesthetically pleasing about having grounds for realism, that is, to having justification for believing that besides accounting for the observed phenomena our theories captured something like the True fabric of the world. Realism may well capture the psychology of many or most scientists investigating natural phenomena. However, that psychological fact does not justify the unwarranted epistemic jump from theory construction to declarations about the structure of the world beyond the phenomena we observe. We are either left with searching for truth (with a lower-case t) where we make statements like, "theory *x* is *more* true than theory *y*," or we have to give up speaking about theories as true altogether.

The historical development of astronomy will serve to exemplify the pitfalls of scientific realism. Following the astronomical work of Eudoxus, who presented his solar system model in instrumentalist terms, Aristotle adopted a very similar mathematical model but changed the metaphysics by making the realist claim that the spheres of planetary trajectories had physical existence (Crowe, 1990). Ptolemy improved and codified the astronomy of the ancient world in the *Almagest*. Ptolemy demonstrated that his two different solar system models, the eccentric and the epicyclical, were mathematically equivalent; that is, they both "save the appearances" equally accurately.[2] For several reasons his geocentric, epicyclical model was taken to be the real motion of the heavens in western Christendom through the 16th century.

Ptolemy's model predicted the appearances at least as well as, if not better than, Copernicus' original heliocentric model, and it had the added advantage that it made common sense; the earth just does not seem to be moving. The advantage to Copernicus' model was not that it made better predictions but rather that it seemed less *ad hoc*. This shows that even theories that seem the most unshakable are susceptible to repudiation. The point is, realism is unjustified because from the perspective of any theory it is not possible to say whether or not a better theory or a different theory will supplant it. It is wise, therefore, not to make pronouncements about that for which there is in principle no way to judge.[3]

Newton's *Principia* codified modern mechanics and astronomy. Once again, the picture painted by Newton's work became solidified and thought of in realist terms.[4] When Einstein overturned Newtonian mechanics, one of the insights gained was of not defining physical absolutes. One of the problems with Newtonian mechanics comes from Newton's definitions of absolute space and absolute time. These definitions express metaphysical beliefs about the structure of the world, which Newton could not support with his empirical method. In fact, they are unsupportable. Percy Williams Bridgman (1927) calls Einstein's insight about these and other physical quantities, such as simultaneity, operationalism. Operationalism is the practice of defining things in terms of the process used to measure them. Bridgman claims that this is the only way to avoid having to reject and replace whole theories because of false assumptions. This practice does not ensure that theories will not be overturned when more empirical evidence justifies a change; it just helps keep the concepts in check with what is empirically justified. This example further illustrates the lesson that the history of physics has taught us repeatedly, that our scientific epistemology does not warrant taking a realist position.

Still more problems for scientific realism are revealed by cases of competing, incompatible theories, which are independently confirmable. Einstein's theory of Special Relativity, a deterministic theory, has been empirically confirmed. Within Einstein's lifetime, Quantum Mechanics, a probabilistic theory had also been empirically confirmed. These two empirically confirmed mechanical theories are in this sense incompatible. Einstein failed to accept quantum mechanics, famously saying that he "could not believe that God played dice with the universe." Einstein's realist belief prevented him from accepting what physicists now think of as a basic fact about the world, namely, the validity of Quantum Mechanics.

It might well be the case that string theory, which is only now becoming empirically testable (for instance with the new CERN accelerator which came on-line in 2008),[5] may reveal that we have been considering the problem from the wrong angle, and again overturn our previous best theories. Newtonian mechanics is in a sense wrong and yet although we know it is an approximation, we still use it for a range of phenomena. We continue to use both Special Relativity and Quantum Mechanics despite their incompatibility. There are many examples such as these in nature and natural science. Scientific theory building is not justifiably about Truth: it is about usefulness and accuracy with available data.[6]

1.2 Natural Science as "Best-Bet" for Moral Norms

One could argue that scientific investigation, despite its apparent unsuitability to investigating ethics, is still our best bet for determining ethical norms. Why not put up with the problem of induction, and the necessity of some form of anti-realism like scientific instrumentalism? This arrangement seems to be working very well for the natural sciences. The idea is that even if it does not give the kind of ethical theory we have come to expect from the history of philosophy, that alone should not rule out naturalism. In "The Nature of Mind," David Armstrong argues that philosophically generated theories of mind should defer to whatever the natural sciences have to say about mind. To support this, Armstrong points out that it is only in the natural sciences that consensus is ever reached. Armstrong claims that, unlike what is the case in philosophy, theology, and the rest of the humanities, natural science makes progress insofar as all of its members come to agree about certain theories. Virtually no biologists argue that evolution is not the best explanation for the phenomena of speciation, and extinction events. It seems rather unlikely to come across a chemist who will deny the first law of thermodynamics or that water is H_2O, or a physicist who will argue against the attractive force massive bodies have on each other. Of course there are many, many disagreements, but there are fundamental agreements. Armstrong claims the same cannot be said of other disciplines.

In "Atheism and Theism," J.J.C. Smart takes up this line of reasoning and applies it to naturalism. Smart argues that we should be satisfied to take whatever is the state of the art of the relevant sciences and use this to develop our ethical principles knowing that they might change. After all, this is what scientists themselves do. Theories only last as long as they are useful. This position would ask, "What better can we do?"

Unlike its effect in the sciences, this approach would be giving up too much in ethics. A minimum condition for ethics should be internal consistency. Scientific theories also require internal consistency, but (as shown above) there is not necessarily a problem if there are equally verifiable incompatible theories within a science. This would, however, be a catastrophic problem for an ethical system. Multiple incompatible ethical theories is the problem we have now, and the reason for continued study. Following our scientific methods, we might come up with contradictory ethical theories equally well supported by the data.

1.3 Open Question Argument Reconsidered

The greatest problem with using scientific investigation to determine ethical theories goes beyond the obstacle of contradictions due to inherent relativism. Whether or not there are theory-neutral observations, natural science aims at a kind of objectivity, for instance with reproducibility of experimental results. This is strongly related to Moore's Open Question argument against naturalism. There have been several well known objections to the Open Question argument. However, at the heart of the argument is a claim about basic scientific methodology. The very way that we investigate nature necessitates that we cannot discover a judgment, only data or facts. We cannot discover goodness itself; we can only discover material bodies and relationships, which we then judge to be good, bad, or morally neutral.

I can understand the appeal to want to use natural objects because of the more likely chance of agreement. However, it does not seem that anything could be discovered which would be convincing as a normative first principle. Unlike physics and chemistry there are relatively few things in biology that take on the status of "law of nature" (such as conservation of energy or the first law of thermodynamics).[7] Perhaps it is because models in biology are different from models in the physical sciences.[8] But suppose we did find such a biological law of nature which was that all living things do everything to survive, or that everything seeks pleasure and avoids pain at all costs (this is just what John Stuart Mill relies on in *Utilitarianism*). Mill claims that this is natural for living things, so it makes sense to pursue in a deliberate manner what we pursue by nature. We will see below that something being natural to us does not necessitate it being for the best. Even if we found something like a natural law, we would have to evaluate whether or not we think it is good and should be followed or bad and should be challenged.

To combat the problem that our epistemology is inadequate for establishing the levels of certainty about the world desirable for naturalistic ethical theories, philosophers have put forward ideas such as naturalistic epistemology and evolutionary epistemology. Aristotle's epistemology relies on our human acumen at recognizing and categorizing natural kinds.[9] Aristotle does not demonstrate this theory, it is just one of the things it means to be human; namely, to be good at making generalizations from limited experiences. This unjustified epistemology was unacceptable to modern philosophers, and their responses are typified by Kant's criticism of what he views as doing science by haphazard observations. Kant has in mind such forerunners as Galileo, Torricelli, and Stahl when he praises the new methods of scientific investigation.

> They learned that reason has insight only into that which it produces after a plan of its own, and that it must not allow itself to be kept, as it were, in nature's leading-strings, but must itself show the way with principles of judgment based upon fixed laws, constraining nature to give answer to questions of reason's own determining. Accidental observations, made in obedience to no previously thought-out plan, can never be made to yield a necessary law, which alone reason is concerned to discover (1929 [1787], p. 20).

Kant is criticizing the Aristotelian notion of observation as being incomplete: one can never be sure that one has observed the whole phenomenon, and so Aristotle is never justified in claiming knowledge about natural kinds.

In "Natural Kinds" (1969) Quine takes up the natural kinds charge, defending the position that humans are by nature good at learning about the world. Quine uses as evidence the scientific understanding Aristotle did not have available to him. Quine argues that "[c]reatures inveterately wrong in their inductions have a pathetic but praiseworthy tendency to die before reproducing their kind" (p. 126). Quine is using the fact of our existence, together with evolutionary theory, to infer that we must be pretty good at matching our concepts of the world to the actual world or we would not still be here. Greater detail will be given below explaining the fact that evolutionary forces do not necessarily tend toward perfection, betterment, or epistemic certainty. However, it will be helpful here to discuss some immediate problems with Quine's argument. Steven Stich (1990) goes a long way to dispelling the Quinean application of evolution to epistemology. For instance, Stich points out the potentially very significant difference, evolutionarily, between a trait that leads to "false positives" as opposed to "false negatives" when choosing among potentially poisonous substances. The cost to fitness of avoiding false positives, those substances deemed dangerous that are not actually dangerous, may be comparatively much lower than the cost to fitness of not avoiding a false negative, that is, a substance that is dangerous but that was deemed not dangerous. This fact about evolution explains how unreliable inferential processes might evolve. In fact, there might even be fitness advantages to certain unreal or false perceptions.

To take a non-biological example, consider the gas gauge in your car.[10] An optimistic gas gauge, one that registers more fuel than is in the tank, will likely lead to long walks to the gas station. The analogous relation in nature, as Quine puts it, might, in "praiseworthy" fashion tend to die out. A perfectly accurate gas gauge, on the other hand, one that precisely measures the amount of fuel left in the tank, might be satisfactory for preventing one from running out of gas, if one is sufficiently conscientious. However, a pessimistic gas gauge, one that underestimates the amount of gas in the tank, might in fact be the best at preventing one from running out of gas. In the analogous relation in nature, this would correlate with the highest fitness even though it gives a less accurate depiction of the world than a less fit trait. Hence, Quine's argument for naturalistic epistemology does not go through.

Another attempt at naturalizing epistemology is offered by Hilary Kornblith (1993). Unlike Quine, who uses evolution as evidence that there is a good fit between our concepts of the world and the world itself, Kornblith claims that good fit between our psychological processes and the structure of the world is a fact, and that it is established independently from evolution. Kornblith says that the role of evolution is to explain *why* there is a good fit between our beliefs and the structure of the world after we have established the fact. So, Kornblith avoids the pitfalls of Quine's position. The sticking point for Kornblith is instead to justify the claim that our concepts do line up with the structure of the world so well. As evidence, Kornblith uses the success of science.

Kornblith's argument is stronger than Quine's, but it still does not get us the epistemic certainty needed for grounding moral naturalism. In essence, Kornblith's claim falls prey to the same criticisms leveled against Aristotle's epistemology. The argument is that we know we are good at generating accurate beliefs about the world because our science is successful. If this argument goes through, the implication will be that we must be generally good at learning about the world; however, that generality cannot be applied as justification for any particular belief. As discussed earlier, any scientific theory is subject to revision or abandonment. That many theories have been successful does not alter this circumstance. If we construct an ethical theory based on scientific observations, it might be successful as many theories have been, or it might not be. The fact of general success could not ground any particular observation or theory. Any set of beliefs might be like the accurate gas gauge or the pessimistic gas gauge, or in fact, like the optimistic gas gauge, just not so inaccurate as to have made a significant negative impact.

2 Can We Learn to Be Moral by Following Nature's Lead?

Our contemporary approach to studying the natural world, either by design or by historical accident, is set up to ask empirical questions and test hypotheses by observing material consequences of experiments. As discussed above, if we use our scientific investigation of nature to come up with ethical principles, they either have to be directly observable phenomena or they have to be derived from the phenomena that serve as natural examples of what is moral. We have already seen that we will not be able to use scientific methods to discover the good in nature. The strongest use of nature for the second approach would derive moral norms from natural examples. However, there is a conceptual difficulty that must be addressed by proponents of taking natural facts to directly generate examples of moral facts.

2.1 The Problem with Generating Moral Norms from Natural Examples

Aristotle's virtue ethics in the *Nicomachean Ethics* can be viewed as a kind of naturalism depending on what it is that makes the wise and courageous actions wise and courageous—are they necessary and natural or are they so by human convention? The relevance to this discussion is the fact that the path to finding out what is ethical in Aristotle's system is practicing ethical behavior. "For the things we have to learn before we can do them we learn by doing them" (1998, pp. 28–29). This seems to make sense for house builders, his first example; many professions use apprenticeships as a teaching tool. However, one already has to know enough to be able to identify who the house builders are before one can begin apprenticing. It appears to be much harder to identify the just citizens than it is to identify the house builders. Knowing who to imitate would seem to constitute a fair amount of knowledge of the virtues.

In the *Groundwork*, Kant criticizes Aristotle's ethical theory by pointing out that for someone to be an exemplar of a virtue—temperance, for instance—we first have to know what temperance is so that we can pick out the temperate people, or else we could not know who to imitate. Kant's objection to Aristotle gets at precisely why we cannot use examples from nature to generate our moral norms: we first have to judge which things are in fact examples of good things, which requires that we already have an idea of the good by which to judge natural phenomena. Using natural examples after we have established moral norms might serve a mnemonic or pedagogic purpose, but this entails that we are not deriving our norms from nature.

Once again we see that we cannot derive moral norms directly from nature. At this point, however, there is still room for a weaker form of naturalism that begins with a general first principle and then relies on nature to fill in the moral directives. Such a general first principle might be that the good is thriving, for instance, and that thriving is done best by nature. So, this would lead to attempting to imitate nature because of the belief that nature optimizes organisms or does things for the best.

2.2 *Assuming that What Is Natural Is for the Best Is a Misunderstanding of Biological Processes*

To just follow "nature's lead" as it were, to pick up things common in nature for ourselves we would still have to evaluate each item for whether or not it is good. For instance, we would not want to start eating our young, even though cannibalism, even cannibalism of one's offspring, is perfectly natural, at least in the sense that it occurs commonly in the animal kingdom. It does not seem to be natural for our species, however. So the question at this level might be, what is natural for our species? One answer to this question might be all the things that we do. Could this simplistic answer be sufficient? It is, after all, the way that we determine what is natural for other species: we observe their behaviors. In humans we observe all sorts of behaviors; it is unpleasant to think that activities we now consider to be elements of a life that is "nasty, brutish and short," might have the same moral status as other activities simply because they are observed human behavior. Observing humans reveals all sorts of behaviors, but we still want to call some of them good and some of them bad. Indeed, much of ethics is precisely about constraining natural behaviors. Murder, rape, adultery, and theft, for example, are so ubiquitous in human society that many scientists have attempted to give natural selection accounts of these phenomena. These biological accounts focus on acquisition and protection of resources, especially genetic resources. The range of these explanations has been from Dawkins's (1976) selfish gene theory to evolutionary psychology's controversial attempt to explain sexual aggression (Thornhill and Palmer, 2000). An ethical system that would allow us to judge certain activities to be immoral, even though they may be natural, seems to be an essential element of ethical theories.

This highlights an important difference between the rest of the animal kingdom and the human animal. We are able to judge which of our natural endowments and tendencies we want to support and which ones we want to suppress or eliminate. This does not seem to be in the power of other species. Whether or not we think that it is good that we have this power, the fact of it implies that even following whatever is natural is as much of a choice as any other path. We do not have to limit ourselves to what we decide is natural for us. Instead, we evaluate each behavior by comparing it to something else that we have already decided is good. We ask how similar given behaviors are to a set of principles, or, perhaps more commonly, we ask whether or not the behaviors make us better or worse, bring us closer to our goals or take us farther away.

Here the naturalist will claim that our goals are inherently natural, are determined by the nature of the species. It is problematic, however, to try to insist on what other people's goals are, or should be, by positing that we all have the same goals insofar as we are members of the same species. Nature, as understood through biology, has no goals for us. Leon Kass (1985) argues that living things do have a telos in an individual sense, and that nature or evolution has had an inevitable telos from the beginning, namely, humankind. Kass argues that the telos of the organism is to survive and reproduce. It is argued that organisms without this telos would go extinct. Kass implies that this telos is natural and that there are natural and unnatural ways for us to go about achieving our telos.

> Thanks to the sexual revolution, we are able to deny in practice, and increasingly in thought, the inherent procreative teleology of sexuality itself. But, if sex has no intrinsic connection to generating babies, babies need have no necessary connection to sex. Thanks to feminism and the gay rights movement, we are increasingly encouraged to treat the natural heterosexual difference and its preeminence as a matter of "cultural construction" (Kass, 1997, p. 18).

If reproduction is our telos it goes against our conscious choices. Meston and Buss (2007) make clear that procreation is not among even the top fifty reasons why humans have sex. This is significant if choice is a desirable condition for moral agency.

Kass (1985) claims there has been a progression throughout evolutionary time toward a telos on a grander scale as well.

> To be sure, 'low' forms persist and flourish, but higher types have successively appeared. ... Higher types in what sense? *Higher in terms of soul.* Somehow, as thousands upon thousands of species came and saw and went under, the processes of evolution not only produced more and more organisms, new species and new forms—or, in other words, produced both more soul and more forms for the expression of soul—but also *higher grades* of soul (Kass, 1985, p. 270, original emphasis).

By "soul" Kass means, "the integrated vital powers of a naturally organic body."

> First, is there a natural term to evolution at least in its tendency of ascent? The answer would appear to be man; at least it is hard to see how it could be anything else. Man is the peak, both in possessing the *highest,* and also in possessing the *complete range* of faculties of soul. Even looking to the future, what could be higher than man? (Kass, 1985, p. 272, original emphasis).

We need not be able to foresee the future of the animal kingdom to know that Kass's conjecture that we are the pinnacle for all time does not square with our understanding of biology. Evolution is the process of the changing of gene frequencies in populations over generations. There is no reason to suppose that this process could stop before the last living thing is extinct. I can understand how someone who is so impressed by the human animal that he cannot imagine something better might be inclined to fear change. The person who cannot imagine any improvement should not try to change anything. All of the others who can imagine things better than they are should aim directly at our betterment.

Kass (1997) identifies the traditional process of sexual reproduction.

> Sexual reproduction—by which I mean the generation of new life from (exactly) two complementary elements, one female, one male, (usually) through coitus—is established (if that is the right term) not by human decision, culture or tradition, but by nature; it is the natural way of all mammalian reproduction (Kass, 1997, p. 21).

Surely, Kass's description of the traditional process of reproduction is correct. Kass immediately adds:

> By nature, each child has two complementary biological progenitors. Each child thus stems from and unites exactly two lineages. In natural generation, moreover, the precise genetic

> constitution of the resulting offspring is determined by a combination of nature and *chance*, not by human design: each human child shares the common natural human species genotype, each child is genetically (equally) kin to each (both) parent(s), yet each child is also genetically unique (Kass, 1997, p. 21, my emphasis).[11]

Kass seems to be assuming that whatever traits or structures we have because of evolution are ideal or for the best. This is particularly interesting given Kass's acknowledgement of the significance of chance in evolutionary processes. Chance plays a large role both evolutionarily and in the development of each organism.

Taking a look at the natural processes responsible for all living things will help explain what is so problematic about mistaking what is natural for what is good. The most famous of the five traditionally recognized forces of evolution is natural selection.[12] The focus here will be on natural selection because of the processes of evolution, natural selection is the one thought to have the best chance at working toward "improvement" in species. Natural selection is not a teleological force; it does not have a goal, it does not aim at doing anything. When we are evaluating whether or not we think it is a good idea to follow along with our natural situation, it would be a good idea to keep in mind that natural selection does not optimize species to their environments. Natural selection is the recognition of a stochastic fact; namely, that in a given population, the organisms that produce more offspring, or more offspring that survive to reproductive age, are more likely to pass on their heritable traits than the organisms that produce fewer offspring. It is not possible to say definitively which traits were selected for and which traits are piggy back traits or simply neutral. In fact, all that can be said of a trait (according to natural selection) is that while a trait persists the only thing that can be known for sure is that it has not been selected against with strong enough pressure to drive it to extinction yet. There is no way to judge with certainty a trait to be a beneficial trait, a negative trait, an adaptation, an exaptation, a piggy back trait, and so forth.

What is more, environments are constantly changing due to atmospheric and terrestrial forces as well as through the impact each species has on its environment.[13] The result is that there cannot even be a final or most fit set of traits overall. The same trait that might make members of a population fit in one season may be detrimental in the next. Natural selection is best described as a satisficing rather than an optimizing force.

Over successive generations, coupled with the other evolutionary forces, organisms tend toward local maxima, not necessarily maximum, or optimal fitness, in a given environment. To further clarify this concept, consider a graph of phenotypic space where the x-axis represents possible traits and the y-axis represents fitness (Fig. 1). Let *A* represent a trait with a given fitness, such as a particular pattern of shapes on the back of a butterfly's wings. The point *S* represents a different pattern with a higher fitness than at *A*. Any non-directed mutation that results in a change in the wing coloration pattern that is within the same trough but below*A*, meaning that it is less fit than *A*, will tend to leave fewer offspring than butterflies with the trait represented by *A*. Likewise any mutations that put a butterfly between *A* and *S* will tend to do better than *A*. Natural selection pressure will tend to favor mutations that lead to the pattern of coloration represented by

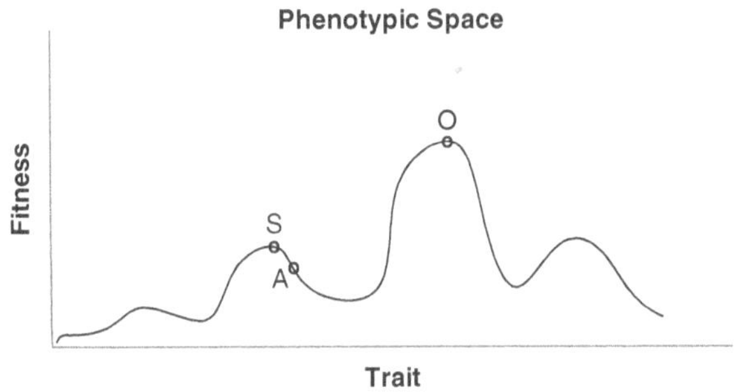

Fig. 1 Natural selection is a satisficing, not an optimizing process. An organism with a heritable phenotype of a particular fitness at *A* cannot be moved by natural selection to the phenotype with the optimum fitness at *O*. Natural selection can only push the organism to the local maximum at *S*

the local maximum of fitness, namely point *S*. Perhaps the trait represented by *A* gives certain camouflage to the butterfly giving it some protection from predators. Perhaps at *S* the pattern fools the butterfly's predators into thinking the butterfly is some other kind of organism that is poisonous to them if eaten. *S* is more fit than *A* because it offers greater protection from predation. But suppose the trait associated with the point *O* is a pattern that resembles the eyes of the animal that preys on the butterfly's predators, which might be the pattern that offers the greatest possible fitness. Natural selection pressures cannot work to move an organism through lower relative fitness, even if the end result will be greater fitness. From the point *A* natural selection could not "push" a species first to less fit traits then to the greater fitness at point *O*. Natural selection is not an agent; it is not calculating, and it will move an organism's traits in the opposite direction of optimal fitness if that is the direction of the local maximum with the greatest relative fitness.

If we take our cues from nature, should we do the same, or should we not aim higher—at optimizing? Optimizing is not what happens by natural processes, but we find that choosing our limits solely according to the satisficing tendencies of nature has undesirable consequences, as the following example shows.

There is a negative correlation between IQ and number of offspring (Retherford and Sewell, 1988). The implication of this correlation is that there is selection pressure against higher IQ scores. Untempered by other factors, this would lower the average IQ. But, generally we think of doing better on IQ tests as being preferable, better, than not scoring as highly. Natural selection is a natural process; if we think that this fact implies that its effects are good, should we help it along? Does this mean that doing things to improve one's IQ score is unnatural and they are wrong?

There is an even stronger negative correlation between level of education and number of offspring, even among socially advantaged families (Scarr and Weinberg, 1978; Retherford and Sewell, 1989). Because there is also a correlation between education levels of parents and offspring, there is then a negative selection pressure, which alone would tend to drive down the level of education of each successive generation.[14] With education, as with IQ scores, we have to ask whether we think the "natural" trend is good or not. Should it be left alone, aided, or resisted? I am happy to resist natural selective pressure against education just as I am happy to wear glasses to compensate for poor vision.

Evolution and natural selection specifically do not lead to optimization. We have our traits by evolution. Therefore, we should not assume that our traits are for the best. People like Kass who argue that we should be more natural, or that following our nature is inherently good, make the mistake of assuming that what is natural is for the best. Of course it would be foolish not to exercise prudence and humility when considering improving on nature, but changing nature cannot be inherently bad because it does not imply change for the worse, because nature does not imply that which is best.

2.3 Less than Entirely Scientific Moral Arguments from Nature

Normative appeal to nature and the natural is not limited to our scientific understanding of nature. It is very difficult to define what is natural, especially, it seems, for our species. Mixed in with scientific knowledge are cultural and personal feelings and views about our place in nature. Thus, when the terms natural and unnatural are used in moral arguments their epistemic origins and meanings are often unclear.

Moral appeal to nature very often is applied only negatively, for instance, when it is argued that *X* is immoral because it is unnatural. John Corvino (1997) takes up the challenge posed by this line of reasoning by examining some of the different ways the term unnatural is used in these kinds of arguments. He finds that the ascription of unnatural to something never necessarily implies that it is immoral. Thus, all arguments of the form unnatural implies immoral are invalid. Although Corvino addresses his discussion specifically to the argument that says that homosexuality is immoral because it is unnatural, his discussion of the uses of the term unnatural applies much more broadly. Corvino points out five separate meanings for the term "unnatural": (i) "what is not practiced by other animals"; (ii) "what is disgusting or offensive"; (iii) "what is unusual or abnormal"; (iv) "what violates an organ's principal purpose"; and (v) "what does not proceed from innate desire".

2.3.1 What Does Not Occur in the Animal Kingdom

This line of reasoning is used to ask questions about specific behaviors and relationships as well as attitudes towards resources. For example, a justification sometimes put forward for a conservationist ethic claims that the human is the only animal that hunts for pleasure. There are two problems with this line of reasoning: (1) the claim that humans are the only animals that hunt for pleasure is misleading, and (2) even if that claim were true, the fact that an activity is unique to humans does not provide sufficient rationale for limiting that behavior.

Misunderstandings about the rest of the animal kingdom often lead to moral condemnation on the false assumptions that only humans for instance, commonly practice homosexuality (Bagemhl, 1999; Sommer and Vasey, 2006), or pose a threat to biodiversity (Dickman, 1996; Biodiversity Group Environment Australia, 1999).

The second sentiment, that we can argue against an activity simply because it is unique to humans, is clearly problematic. Presumably, those who make this kind of argument in print do not think that the uniquely human trait of writing is bad. Since being unique is not enough to condemn an activity, what additionally makes the activity in question immoral? We will see that something extra, besides the behavior's frequency in nature, is always required to make a moral judgment about the behavior. What follows from this is that it is the "something extra" that makes a behavior right or wrong and not the fact that it occurs in the rest of the animal kingdom.

The position being put forward here is not that what we learn about the animal kingdom is irrelevant to humans. However, when it comes to moral judgments, all that new knowledge about non-human animals might do is broach a question about behavior. The fact of the behavior itself cannot serve as a moral judgment.

2.3.2 What is Disgusting or Repulsive

Sometimes when people say something is unnatural they mean that they cannot imagine doing it because it seems disgusting or repulsive. The set of things that people find disgusting includes everything from taboos, such as cannibalism, incest, and necrophilia, to certain diets that include sushi or haggis. Not all of these things involve moral questions. Kass (1997) argues that there is wisdom in repulsion, that what repulses us probably does so because it gets at some underlying immorality. At least this is why Kass thinks certain things disgust us. Kass is arguing against reproductive human cloning (through nuclear somatic transfer). Of course, he does not claim that everything in the set of disgusting things involves moral questions. So, presumably only the moral issues that involve repulsion do so because they reflect a moral principle. Once we see that some moral issues do not involve repulsion and that some repulsive things are not at all moral issues, we see the fallacy in claiming that repulsion is a reliable indication of immorality.

Gregory Pence (1988) points out that in the 1800s when the process of vulcanizing rubber made the manufacture of condoms fast and inexpensive there were cries of unnatural and disgusting at the thought of using condoms. The outcry was strong enough to play a significant role in the passing of the Comstock Law in 1873 which prohibited condom advertisements and distribution of condoms through the postal service. Most people today do not have a reaction of disgust regarding condom use, independently of whether or not they think condoms are morally permissible.

Obviously this is not an argument for the moral permissibility of things that repulse us. The point is that what repulses us and what is immoral are either independent or their correlation is sufficiently mysterious to prevent reliable inference from one to the other. The best we could hope for would be to try to use repulsion as a clue that moral investigation might be warranted. However, every case would have to be investigated to see if what was repulsive was also immoral. Because there is no clear causal connection, it would not be possible to say in advance, or make predictions about, which repulsive things would also be immoral. This is why Kass's argument for the wisdom of repulsion is ultimately unhelpful. Cherry (2005) makes the more universal point about the problem of using emotions in general as a guide for morality:

> If one holds that our natural feelings should be embraced as important, formative elements of morality, then we must have some understanding of the criteria for identifying and ranking the emotions that are to qualify. Do all emotive reactions, by all parties, point to important moral features of a situation? Which emotions should we take as authoritative and how should we quantify them intra- and interpersonally? And if the emotive reactions differ among persons, how can one identify those that, as morally appropriate, ought to guide the formation of public policy versus those that ought to be ignored, if not actively discouraged? Formulating policy utilizing such emotional reactions assimilates and enforces maxims that are very likely merely subjectively contingent; indeed, at times, they may even be pernicious (Cherry, 2005, pp. 40–41).

The point is that the argument that something is repulsive because it is unnatural and immoral because it is repulsive does not go through. It is immaterial whether or not feelings of repulsion indicate that something is natural or unnatural; in either case, the argument fails because repulsion does not entail immorality.

2.3.3 What Is Rare or Unusual Is Unnatural

Another use of the term "unnatural" refers to behavior that is uncommon. In this case, the connection between the terms normal and normative value is seen. The idea is that what is not normal is unnatural. The tricky thing about this argument is first figuring out which things are not normal and then figuring out which things that are not normal are the bad ones. This again shows that the fact of the practice itself does not give us a moral judgment or guideline. There are all sorts of physical traits which occur rarely; some are desirable, many are not. It is hard to imagine that we could determine a standard for exactly how infrequent something had to be in order to qualify as "not normal." Which is rarer, exceptional beauty or exceptional homeliness? Does this make one more unnatural than the other? Does it make one more moral than the other?

An easy answer could point out that traits beyond one's control are not moral issues because no choice is involved. What about cases where someone with a physical deformity, who has the resources to change the deformity, chooses not to? It is not clear whether the more "natural" thing to do would be to restore the deformity to what is normal, or to accept what happened through natural processes even though it is not the norm. There is a raging debate in the deaf community, for example, about hearing restoring procedures. Most hearing people would agree that having their hearing is far preferable to not having their hearing. There are, however, many hearing impaired people whose hearing might be restored with operations which they choose not to undergo. The question is even more controversial when it comes to the operations for their children who might be helped by cochlear implants, or in cases where deaf parents have a hearing child, but want the child to be surgically rendered deaf so that he fits into Deaf culture. These difficult moral questions are not aided by questions about what is normal, common, and natural. The different people involved all have different ideas about what counts as normal, common, and natural. The issue here is that science, or the natural, cannot tell us which types of body image and abilities morally to prefer. The criteria for that judgment must come from elsewhere.

How common does something have to be for it to be normal? We see the difficulty as soon as we start suggesting numbers. Is 95%, 80%, 70% okay? Okay for whom? Living to be 80-years-old used to be very rare, but it would sound strange to call someone's continuing to live unnatural. No one today lives to be 150 years old. Does that mean it is unnatural to work toward that? Is it necessarily a good thing that everyone dies before they are 150 years old?

2.3.4 Unnatural Use of Organs

One argument from unnatural to immoral refers to the way organs are used. The claim is that using an organ in ways other than for the organ's purpose is unnatural. This argument in particular has been most often used against homosexuality. The argument goes that each organ has a purpose or intended function. The genitalia are for reproduction. Therefore, using the genitalia for non-procreative activities is unnatural. Of course this is simultaneously an argument against homosexuality and any sexual act where reproduction is not both plausible and the intended outcome.

This assumes that organs have functions or purposes other than those of the owner's choosing. For that to be the case in a strong sense there would have to be some directing force. For instance, if there is a creator, it might be the case that the creator intended only one or a specific use per organ and that all other uses are unnatural. In this case, we could only know if unnatural implied immoral by knowing the creator's judgment on such matters. This has been somewhat controversial within theology. More importantly, it again requires a second type of judgment (in this case a non-scientific/non-rational one) to know what is good and what is bad.

In the case where creationism is false, we rely on what biology has to tell us about functions. The safest way to discuss functionality in biology is to talk about the relationships among the organs without referring to purpose. Why are purposes problematic? As organisms evolve, so does the way they use their organs. The classic example is the use of feathers by birds for flight. Our best understanding tells us that feathers were adaptations for thermal protection. Much later in evolutionary history, feathers were co-opted for flight. This is called an exaptation. Does this mean that using feathers for flying is unnatural? The human mouth gives us another example. Using the mouth for language came about long after the mouth was used for breathing, eating, and vomiting. The fact that no one claims that speaking is unnatural is evidence that determining the purposes for organs in moral arguments had largely occurred after the ethical judgment had been made.

What about using an organ in a way that only some people do? A common surgery for someone who has lost a thumb is to remove one of their toes and attach it to their hands. This is certainly the sort of thing that strikes some people as unnatural. It is not something that happens without artificial interference, and it uses a part of the body in a way not "intended" by nature. However, for the patients it is a miracle surgery that restores essential functionality. It is very beneficial and probably unnatural, but it would be hard to make the argument that it is immoral.

2.3.5 "What Does Not Proceed from Innate Desire"

Corvino (1997) points out that one argument for what is immoral because it is unnatural claims that behaviors that stem from desires that are not "innate" are unnatural. In the debate about the morality of homosexuality, an important question has been whether people are homosexual by nature, by culture, or by choice. Corvino argues that it is a mistake to think that if someone has homosexual desires by nature that that entails that it is moral to act on those desires. He gives the example that some people seem to have more violent desires than others by nature, but that does not entail that their violent acts are necessarily moral acts. However, if someone has homosexual desires not by nature, this does not imply that their homosexual actions based on those desires are immoral. It is not natural for humans to swim under the water for an hour, but that does not make their scuba diving immoral.

2.4 At a Minimum, Does Nature Present Ethical Boundaries?

So far we have certain boundaries or limits to our investigation of the natural world. Our epistemological range for empirical study seems to be bounded by the speed of light, on the one hand, and Plank's constant, on the other. Does nature likewise give us a set of moral boundaries past which we may not go? There are examples of people interfering with natural processes with catastrophic results. Introducing non-indigenous species in an attempt to control environmental changes sometimes leads to much worse changes. There is concern about genetically modified foods (GMOs) because predicting exactly how they will impact the environment is extremely difficult. There has been virtual panic about the unfamiliar and apparently unnatural processes of nuclear somatic transfer (cloning). The issues are not morally trivial; we are often weighing environmental impact against alleviation of human suffering. So each of these developing technologies must be considered carefully, but nature alone cannot guide us. As we have seen, natural does not imply better and unnatural does not imply worse. Nature cannot provide the moral judgment we need to decide in every case whether to respect or challenge the natural.

3 Conclusion

Perhaps a different scheme for doing science could be constructed that might yield answers to questions about moral norms. It would undoubtedly look very different. It is difficult or impossible to imagine other paradigms for investigating nature replacing ours, but our current empirical methods for investigating the world cannot yield value statements. For now, I will offer that the best we can do is deliberately devise those principles we think best, whether or not they are natural. I hold up John Rawls's (1971) attempt at grounding social institutions morally as a potentially fruitful approach to producing moral norms. Rawls's own conclusions, or perhaps more so the implications of his theory of justice, are certainly not without their problems, but the approach is good. We ask what initial conditions would yield a desirable social structure independently of position. The approach requires only that we figure out what the consequences would be for any given set of starting principles. It does not limit us to whatever the natural sciences might discover. Nothing would prevent us from trying out as a starting point anything a scientist said was a natural principle but we would not be limited to that, and we would not have to worry about forcibly changing systems with new discoveries unless we thought it a good idea. This deliberate approach where we choose what kind of a world we want to live in, is our best option for ethical theory building.

Notes

1. "Every time we proceed to explain some conjectural law or theory by a new conjectural theory of a higher degree of universality, we are discovering more about the world, trying to penetrate deeper into its secrets. And every time we succeed in falsifying a theory of this kind, we make a new important discovery. For these falsifications are most important" (Popper, 1985a, p. 167). In the following quotation Popper compares attaining objective truth to summiting a mountain peak constantly hidden in clouds. "The climber may not merely have difficulties in getting there—he may not know when he gets there, because he may be unable to distinguish, in the clouds, between the main summit and some subsidiary peak. . . . Though it may be impossible for the climber ever to make sure that he has reached the summit, it will often be easy for him to realize that he has not reached it (or not yet reached it); for example, when he is turned back by an overhanging wall. Similarly, there will be cases when we are quite sure that we have not reached the truth. Thus while coherence, or consistency, is no criterion of truth, simply because even demonstrably consistent systems may be false in fact, incoherence or inconsistency do establish falsity; so, if we are lucky, we may discover inconsistencies and use them to establish falsity of some of our theories" (Popper, 1985b, pp. 185–186). Popper is not saying that theories only have value when they are disproved; that is, when they reveal what is not the case about the world. New theories of greater complexity that pass initial tests and displace older theories usually mark some progress. However, he is making the point that epistemic certainty only comes with falsification.
2. The fact that Ptolemy had two different models and the discussion in the first book of the *Almagest* about saving appearances, both constitute some evidence that Ptolemy was a scientific instrumentalist. Despite this evidence, however, a strong case can be made that Ptolemy's ultimate preference for the epicyclic model and his realist language in the later books of the *Almagest* indicate that Ptolemy likely believed that the epicyclic model was true, that it captured the real movement of the eternal and divine celestial spheres.
3. This is the case whether or not we accept Kuhn (1996) paradigm shifts.
4. Newton himself sends mixed signals about the ontology of natural physics. Newton's explicit method is continually to look for the best mathematical model to describe the phenomena. By best, Newton has in mind most accurate fit to the appearances. In Book III of the *Principia*, Newton puts forward the idea that each model or theory should be held only until new data comes to force it to be changed or rejected. When deciding between competing theories, parsimony dictates that we use the simplest complete theory. This is Newton's version of Ockham's razor. Interestingly, the razor can be thought of in realist terms; for instance, by those who think the simplest theory is more likely to be the true theory because simplicity reflects the preferences of the creator, or the razor can be taken as an instrumentalist's paradigm—there are always competing theories, but since we cannot use the

theories themselves to adjudicate which is better, we have to come up with practical conventions for breaking ties. This second view of the razor makes no assumptions about whether or not the universe shows any greater tendency towards more simple or more complex. It is simply more convenient for us to use simpler models when we can, so it makes good pragmatic sense. On the other hand, there is also evidence to suggest that Newton saw himself as writing the new bible, of truly explaining God's work. If this is correct, Newton was a scientific realist. Regardless of Newton's own intentions, however, Newton's worldview became codified and accepted as the true dynamical picture of physics.

5. CERN, the European Organization for Nuclear Research, is home to the new Large Hadron Collider (LHC), which is the world's largest and most powerful particle accelerator. The LHC will be able to simulate the conditions in the first moments after the Big Bang. Scientists at CERN claim they will be able to shed light on previously elusive quandaries; for instance, they expect to be able to perform experiments testing aspects of string theory and to be able to detect the Higgs Boson, if it exists, which may explain the origin of mass.
6. These examples are all from the physical sciences. Looking for natural ethical facts would be more likely to rely on the biological than on the physical sciences. Modern biology is a younger discipline than modern physics. Depending on how you start counting, biology is only as old as either Lamarck introducing the concept of explaining biological phenomena around 1800 (this marked a change away from the older system of merely naming and cataloging organisms, codified by Linnaeus). Or, one could argue that modern biology did not begin until 1859 with the publication of the first edition of *On The Origin of Species*. In any event, we may be in for many more big changes in biology as we have seen in the older science of physics. We continue to use our theories, but it would be a mistake to think that we have reached the end, that we have now truly explained the structure of the world.
7. Sterelny and Griffiths (1999) in *Sex and Death* point out that one such law may be that the members of sexually reproducing species die. The meaning of death is more ambiguous in asexually reproducing species. In *On the Soul* Aristotle takes it as a law of nature that all souled beings strive to participate in the "Eternal and Devine," mainly through reproduction. Humans do this, but also have reason because of our *nous*, which is similar to the *nous* that is the Prime Mover.
8. Keller (2002) explains that, as opposed to other sciences in which models are used to represent classes of phenomena, in developmental biology, models are used to represent classes of organisms. Unlike sciences that desire simplicity above all else in experiments, choosing species for developmental research is particularly complicated because there is inherent conflict between the need for consistency and the need for variation in experimental models. This is generally true in biological experimentation (Rheinberger 1997). This fact about biological research would make finding moral norms in biology even more complicated because organisms that are suited to experimentation are often not the species that is most representative of a class of organisms. Would our ethical norms be derived from the natural or from the natural as affected by captivity effect?
9. This is apparent in the *Metaphysics* in the discussion of *nous* and in the *Posterior Analytics*, especially Book II.
10. Thanks are due for this example to R. J. Hankinson in personal communication, June 2007.
11. Each child is genetically unique; however, identical twins are very similar, more similar genetically, in fact than a clone would be to her parent.
12. Five evolutionary forces lead to change in genetic variation within populations: (1) Mutation: the source of all variation (mutations are undirected with respect to extrinsic goals); (2) Natural Selection: can increase or reduce variation; (3) Gene Flow: moves variation from population to population; (4) Genetic Drift: changes population due to random chance; (5) Catastrophes: extreme environmental events isolate and remove whole populations.
13. Many theorists believe that the earth's atmosphere used to be a reducing one, which allowed for initial life, and then the early life forms changed the atmosphere into an oxidizing one, making more complex life forms, such as us, possible. If this is true, the single greatest change to the earth's atmosphere since the appearance of life was not caused by humans.
14. In this case, especially, there seem to be offsetting factors.

Bibliography

Aristotle (1984a). *Metaphysics*, W.D. Ross (trans.), in J. Barnes (ed.) *The complete works of Aristotle*. Princeton: Princeton University Press.
Aristotle (1984b). *On the soul*, J.A. Smith (trans.), in J. Barnes (ed.) *The complete works of Aristotle*. Princeton: Princeton University Press.
Aristotle (1993). *Posterior analytics*, J. Barnes (trans.). Oxford: Clarendon Press.
Aristotle (1998 [c. 335 B.C.]). *The nicomachean ethics*, D. Ross (trans.). Oxford: Oxford University Press.
Armstrong, D. (1970). The nature of mind, in C.V. Borst (ed.), *Mind/brain identity*. London: Macmillan, 67–79.
Bagemhl, B. (1999). *Biological exuberance*. New York: St. Martin's Press.
Biodiversity Group Environment Australia (1999). Threat abatement plan for predation by feral cats. [On-line]. Available: www.environment.gov.au/ biodiversity/ threatened/ publications/tap/cats/index.html.
Boyle, J. (1999). Natural law and global ethics, in M.J. Cherry (ed.), *Natural law and the possibility of global ethics*. Dordrecht: Kluwer Academic Publishers.
Bridgman, P.W. (1927). *The logic of modern physics*. New York: Macmillan.
Cherry, M.J. (2005). *Kidney for sale by owner*. Washington, D.C.: Georgetown University Press.
Corvino, J. (1997). Why shouldn't Tommy and Jim have sex?, in J. Corvine (ed.), *Same sex*. Lanham: Rowan & Littlefield.
Crowe, M.J. (1990). *Theories of the world from antiquity to the Copernican revolution*. New York: Dover.
Dawkins, R. (1976). *The selfish gene*. Oxford: Oxford University Press.
Dickman, C.R. (1996). *Overview of the impact of feral cats on Australian native fauna*. Report to Australian Nature Conservation Agency.
Gettier, E. (1963). Is justified true belief knowledge? *Analysis*, 23, 121–123.
Kant, I. (1981 [1975]). *Grounding for the metaphysics of morals*, J.W. Ellington (trans.), Boston: Hackett Publishing.
Kant, I. (1929 [1781]). *Critique of pure reason*. N.K. Smith (trans.). New York: St. Martin's Press.
Kass, L.R. (1985). *Toward a more natural science*. New York: The Free Press.
Kass, L.R. (1997). The wisdom of repugnance. *New Republic*, 216(22), 17–26.
Keller, E.F. (2002). *Making sense of life*. Cambridge: Harvard University Press.
Kornblith, H. (1993). *Inductive inference and its natural ground*. Cambridge: MIT Press.
Kuhn, T.S. (1996). *The structure of scientific revolutions*, 3rd ed. Chicago: University of Chicago Press.
Lenman, J. (2006). Moral naturalism, *The Stanford encyclopedia of philosophy* (Fall 2006 Edition), E.N. Zalta (ed.), [On-line]. Available: http://www.plato.stanford.edu/archives/fall2006/entries/naturalism-moral/.
Meston, C. and Buss, D.M. (2007). Why humans have sex. *Archives of Sexual Behavior*, 36, 477–507.
Mill, J.S. (2002 [1861]). *Utilitarianism*. Boston: Hackett Publishing.
Moore, G.E. (1903). *Principia ethica*. Cambridge: Cambridge University Press, 1903; Revised edition with "Preface to second edition" and other papers, T. Baldwin (ed.), Cambridge: Cambridge University Press, 1993 (page references in this entry are to the revised edition).
Pence, G.E. (1988). *Who's afraid of human cloning?* Lanham: Rowman & Littlefield.
Popper, K. (1985a). The aim of science (1957), in D. Miller (ed.), *Popper selections* (pp. 162–170). Princeton: Princeton University Press.
Popper, K. (1985b). Truth and approximation to truth (1960), in D. Miller (ed.), *Popper selections* (pp. 181–198). Princeton, Princeton University Press.
Ptolemy (1948). *Almagest*, R.C. Taliaferro (trans.), in *Great books of the western world*, Vol. XVI. Chicago: Encyclopedia Britannica.

Quine, W.V.O. (1969). Natural kinds, in *Ontological relativity and other essays*. New York: Columbia University Press. pp. 114–138.

Rawls, J. (1971). *A theory of justice*. Cambridge: Harvard University Press.

Retherford, R.D. and Sewell, W.H. (1988). Intelligence and family size reconsidered. *Social Biology*, 35, 1–40.

Retherford, R.D. and Sewell, W.H. (1989). How intelligence affects fertility. *Intelligence*, 13, 169–185.

Rheinberger, H.-J.. (1997). *Toward a history of epistemic things: Synthesizing proteins in the test tube.* Stanford: Stanford University Press.

Scarr, S. and Weinberg, R.A. (1978). The influence of 'family background' on intellectual attainment. *American Sociological Review*, 43(5), (Oct), 674–692.

Smart, J.J.C. (1996). Atheism and theism, in J.J.C. Smart & J. Haldane (eds.), *Atheism and theism.* Oxford: Blackwell, 6–83.

Sommer, V. and Vasey, P.L. (eds.) (2006). *Homosexual behaviour in animals*. Cambridge: Cambridge University Press.

Sterelny, K. and Griffiths, P. (1999). *Sex and death*. Chicago: University of Chicago Press.

Stich, S. (1990). *The fragmentation of reason.* Cambridge: MIT Press.

Thornhill, R. and Palmer C.T. (2000). *A natural history of rape*. Cambridge: MIT Press.

Moral Acquaintances and Natural Facts in the Darwinian Age

Stephen S. Hanson

Standard interpretations of the relevance of natural facts through natural law theory tend to fail once one considers the full implications of evolutionary theory. This chapter argues that natural facts about human beings can still have moral meaning in this context. After a brief discussion of natural law, it is shown that the moral acquaintanceships between human beings (and possibly other beings as well) can give moral meaning to many of the facts about human existence. The importance of these facts may appear similar to the importance granted by natural law theories—and the relevant facts themselves may be quite similar—but the systems differ both in their justification for the importance of natural facts and in the relevance of contingently true facts.

1 Introduction

In his *Natural Theologies* (1802), William Paley presents a thought experiment wherein a traveler on a heath encounters first a stone, and then a watch, on the ground. The traveler infers from the complex design of the watch that it could not have naturally occurred, but rather must have been created. Consequently, some creator of that watch must exist.

Paley goes on to make a comparable argument for the existence of a divine creator with regard to the complexity in the existence of the world as it is, and life and the features of nature—he notes at one point that the eye compares well to a telescope, and in fact displays more complexity (Paley, 1802, Chapter III). As he wrote before the theory of evolution was developed, his argument does not work as it stands, for it depends too much upon there being no alternative explanation of the existence and development of complex natural features. Evolution provides such an alternative explanation. But one may also employ his example to make a separate

S.S. Hanson (✉)
Department of Philosophy and Family and Geriatric Medicine, University of Louisville, Louisville, KY, USA
e-mail: stephen.hanson@louisville.edu

M.J. Cherry (ed.), *The Normativity of the Natural,* Philosophical Studies in Contemporary Culture 16, DOI 10.1007/978-90-481-2301-8_13,
© Springer Science+Business Media B.V. 2009

but related point. Imagine that the traveler on the heath has never before seen a watch. He might still infer that the watch was created; he could also reasonably assume that it had some purpose: if someone has taken the time and effort to create something intricate and complex, there probably was a reason to have built it. He could infer not only a designer, but a design, from his discovery and examination of the complex watch. The hypothetical traveler could even discover what that purpose was through careful analysis of the object—observing, perhaps, that the arrows on the face always point in a certain direction at the time when the sun is highest in the sky, noting that the hands move at regular intervals and ratios, and so forth—and careful thought about what its function might be based on those observed facts.

I believe that it is not an unfair comparison to suggest that this latter practice is akin to what most natural law theorists seek to do to develop moral theory. These theorists seek to discover the function of humanity, or of humans as individuals, and how best to attain that function. In essence, they argue, we discover ourselves in existence, and wonder what the purpose is for that existence. Through discovery and analysis of natural facts, and careful consideration of those facts, one may determine that purpose. Natural facts are, therefore, an important part of understanding the role and function of human existence. Like the traveler examining the watch, an examination of these facts is how one would discover that function; thus, natural facts are crucially important for this process.

However, prior to even beginning the process, one must first believe that a function is there to be found; otherwise, there would be no point to searching for it. Recall that Paley's traveler first encounters a stone. That stone has been rendered in its present form and location by millions of interactions over a very long time, including some or all of things such as the composition of the Earth at its creation from a conglomeration of dust and gas, formation of the particular type of rock over millions of years by pressure, tectonic motion cracking a large section of stone and exposing a previously buried layer, a portion of that stone's being shattered during use as a catapult stone, long-term erosion of the shattered remains by wind and water, smaller parts being thrown about by young children, and so on. Yet, though it has been created in its present form by all these events, and had one or more of them not come to pass it would not be there in its present form, the traveler does not, and ought not, conclude from this that the stone was designed in any intentional way, nor that it has any particular purpose. Even were it now to be just the right size and shape for skipping across a pond, there is no reason to think it was designed to be that way rather than just becoming that way by happenstance. Consequently, Paley and his hypothetical traveler spend no time examining the creation and function of that stone.

The dichotomy here drawn between the watch and the stone is intentional. Natural law seeks to determine the good of man, a *telos* for human existence, and from that determine what persons ought to do, how they should live, and so on. Traditionally, this treats humans as if they had come into existence in the fashion of the watch, rather than in the fashion of the stone. Like the watch, that there is a designer makes it possible, and indeed likely, that there is a function for which human beings were created. Whereas the examination of a watch can determine the function that its designer had in creating it, natural law theories seek the purpose that the designer of humans had in creating humans. If a deity has created humans into a particular shape, form and function, determining the way that humans were created to function will allow understanding of the human *telos*. It would be reasonable to assume that one could derive the whole, or at least a significant portion of, morality from the exploration of and careful thought about natural facts of human existence.

The existence of a divine creator also provides a clear answer (though not the only possible answer) to Glaucon's challenge of "Why ought one to behave morally?" (Plato, *Republic*, Bk. II). If morality is derived from the purpose for which we are created, then the answer is simple: we should act morally because that is what we were created and designed to do. Thomas Aquinas explains this point by recognizing the natural law as part of the eternal law of all of creation. Mark Murphy notes that "[t]he eternal law, for Aquinas, is that rational plan by which all creation is ordered (*Summa Theologica*, IaIIae 91, 1); the natural law is the way that the human being 'participates' in the eternal law (*Summa Theologica,* IaIIae 91, 2)" (Murphy, 2002). With a divine creator placing humans within a complete creation, with their role to play in it, it is easy to see why it is right to follow that plan. For Aquinas, natural law is simply the basic principles of practical rationality (Aquinas, *Summa Theologica*, IaIIae 94, 2; Murphy, 2002), so it may even be irrational to go against these principles. All of this comes from the understanding of humans as created in their present form, intentionally, by some creator.

If, on the other hand, we have evolved to our current form merely by many millions of small interactions due to the mechanisms of evolution,[1] without any particular guiding hand aiming at a particular form and function, then there is less reason to think that we can derive the whole of morality from the natural facts about our existence. Since evolution is not a purely chance-driven process, this is not exactly like the way that Paley's stone came to be, but it is similar in the relevant way that there is no specific intentional form and design of humans. Our form is what it is because that has been a useful form in the environments in which we developed. There is no function for which we were intentionally created; there is no *telos* for which we were planned (Dawkins, 1996; Gould, 1997).[2]

It is important to understand this point, as it is not always recognized. Evolutionary theory has been dogged with errant reasoning since it began; one of the more pernicious of these errors is an assumption of "progress" and what might be called "laws of structural determination" (Kerr, 2002, p. 333). Evolutionary change is a process that provides potentialities and limits, but it does not follow any predetermined or predeterminable path, nor does it reflect any prior laws (Gould, 1997, 1981). This includes the frequently assumed law that evolution inevitably leads to greater complexity, as well as the idea that evolution leads to ever more perfect and complete beings. Understood that way, humans could be seen as the pinnacle towards which all evolution (so far) has been aiming; one could derive general structural laws and thus a function. But this is an erroneous viewpoint. Evolution is directionless; different species successfully fit into various ecological niches or they fail. We have a role that we play in our environments and in the continuum of living things, but it is not the role we were *meant to* play, or one that was predetermined by structural laws. It is just the role we actually are playing. It just *is*, and no more.

Thus it appears to be very important to the normal function of a natural law moral argument that there be a creator of some sort. The existence of the creator ensures that the *telos* of humankind exists, and that it is actually relevant to morality, that is, why it is morally right to pursue that *telos*. Without the creator, the importance of the natural facts which are supposed to determine for us the purpose for our existence loses its grounding. If the natural features of our lives are that way simply because they have developed that way to fit environments in which our ancestors lived, then there is no particular reason why they ought to form the basis of the whole of our moral understanding.

2 Ancient Origins of Natural Law: Function without a Deity

An initial response could be to argue that one can seek the function of humanity without any explicit reference to a deity. Aristotle's account of the nature of man in Book I of the *Nicomachean Ethics* is non-theistic, and from it he derives both a *telos* for humans and a complex moral theory. Moreover, he does so from an exploration of natural facts and a careful thoughtful analysis of what the meaning of those facts might be. In other words, without any theistic references he accomplishes the goal of natural law of employing natural facts to find a *telos* for man and employs that *telos* to describe morality.

Upon further examination, I believe that Aristotle's arguments cannot provide a function of humans or, if they do, that the function they provide cannot ground morality. That he discovers facts that seem to indicate a function is not surprising. Humans, having a strong tendency to seek patterns (Ackerman, 2004), tend to find them where none actually exist;[3] which is to say, we seek ways to order and understand logically even that which is not actually ordered or logical. There are many natural features that humans have that fit them well into the many niches they occupy in the environment; like Paley, Aristotle did not have evolution as a plausible alternative theory to explain why that could be. Perceived under these circumstances, the obvious good logical fit of these many natural facts does tend to lend itself to inferring and seeking a function. It does not follow from this that man necessarily has a function. Absent an understanding of how else it could have come to be, one could conclude that the stone, which is just the perfect size for skipping across the water, was designed for that purpose, when in fact it merely meets those criteria by accident. If humans have the features that appear as a function only by happenstance, the fact that they can be organized and perceived as a function does not make it so.

Consequently, one must assess not the content of the function Aristotle derived, but rather his argument for the existence of a human function. He sketches a two-part argument for the existence of a human *telos*. First, he argues that the various parts of the human body have a function—eyes, hands, feet, all have functions (Aristotle, 1908, Bk I.7). If a part of an object has a function, the whole must have a function to which the function of the part contributes (Aristotle, 1908, Bk I.7); so the whole (human) must have a function. Second, he argues that the function of humans must be that which is essential to being human; that is, whatever it is that uniquely makes us human, rather than something else, must be our unique function. It is from this that he derives the function of humans as "activity of soul which follows or implies a rational principle. . ." which means activity of the soul in accordance with the best and most complete virtue (Aristotle, 1908, Bk I.7). Fewer modern scholars embrace the pure essentialism of the second step, but the former argument may still initially seem compelling. Parts of the body seem clearly to have functions—the eye to see, the heart to pump blood, and so on—and it is not implausible that these functions must contribute to a whole that also has a function. Otherwise, the function of the parts would seem to exist for the good of themselves, but not for the good of the thing of which they are a part. Indeed, one can hardly speak of the function of a part of the body without reference to the whole:

> We may remark, then, that every virtue or excellence both brings into good condition the thing of which it is the excellence and makes the work of that thing be done well; e.g. the excellence of the eye makes both the eye and its work good; for it is by the excellence of the eye *that we see well* (Aristotle, 1908, Bk.II, 6, emphasis added).

From this, Aristotle concludes, and modern reasoners might also conclude with him, that there must be a proper function of humans.

Once the theory of evolution has been introduced, however, this line of argument seems far less likely to prove a *telos* of the whole, at least in a form useful for moral theory through natural law. Eyes, hands with thumbs, and other body parts were retained by humanity's ancestors because they were (and still are) useful in some sense for genetic reproductive success, whether through better survival rates, better reproductive ability, or other means. These parts, and their precursors, developed by mutation and are employed by the beings possessing them, not because they developed for a particular use but because a particular use of a variation turned out to be helpful. Human great toes are very helpful in walking upright and balancing, but they were not developed to assist in walking; rather, beings with stubby and less oppositionally placed great toes are better able to walk upright. Conversely, those with more offset and thumb-like great toes are better at grasping things with their feet. Both sorts of beings can do the opposite action—humans can grasp with their toes, and chimpanzees and gorillas can walk upright—but one arrangement of the "thumb" toe works better for each action. Both developed over time from a similar part (the "thumb" of the foot) and were used by different beings in different ways; neither has a particular function but rather a use to which it is put.

Consequently, Aristotle's antecedent may not be correct. Eyes, great toes, and the like may not have functions. They appear to, and prior to an understanding of how beings could develop over time to fit a niche, and how those that did not fit that niche would tend to die off over time, describing these uses as functions may have seemed appropriate. But after evolutionary theory provided a possible alternate explanation, mere semblance of an intentional function is not enough to ensure it. If the antecedent is false, the argument fails.

One may, however, read these incidentally developed features as having functions not by intentional creation, necessarily, but merely by their having been adopted as a means of performing a particular function. Great toes have no originally intended function, but it is not a great stretch to say that they have one now, simply because of the way that humans as a rule use them. Such a reading will allow that parts of the body have a function, and this would allow Aristotle's argument to entail a non-intended but still existent function of the whole as well. Further, there does seem to be a function of humans consistent with evolutionary theory, not requiring any presumption of divine design, and which could be described as a law of nature; the evolved functions of parts do contribute to this function. Though this result may initially seem compelling, it will not suffice for a moral theory.

The function in question comes from the way evolution operates at a micro-level. According to one theory of how evolution operates, human beings exist largely as a means of, and thus in a sense for the function of, propagating all or some part of their genes (Dawkins, 2006). The function of humans would then be understood as something like, "successful propagation of (a significant portion of) one's genetic material into the next generation."[4] This is consistent with Aristotle's criterion for a function as described in NE, Bk I.7:

> Let us again return to the good we are seeking, and ask what it can be. It seems different in different actions and arts; it is different in medicine, in strategy, and in the other arts likewise. What then is the good of each? Surely that for whose sake everything else is done. In medicine this is health, in strategy victory, in architecture a house, in any other sphere something else, and in every action and pursuit the end; for it is for the sake of this that all men do whatever else they do. Therefore, if there is an end for all that we do, this will be the good achievable by action, and if there are more than one, these will be the goods achievable by action.

This possible function also fits in well with Aristotle's argument about the function of the parts necessitating a function of the whole. If one were to argue that various parts of the body clearly do have functions, even if those parts were simply developed over time and their functions "discovered" *ex post facto*, the purpose for which all of these functions exist is to better promote the successful propagation of one's genetic material into a future generation. This seems reasonable—if one has eyes, one is better able to see danger or food, and to know the difference between the two, and thus better able to survive and reproduce. Similar stories can be told about other body parts. The function of each contributes to the proposed function of the whole.

Nevertheless, the function of humanity as being propagation of genetic material will not serve well as a basis for a natural law moral theory. It should not and does not follow from that presumed function that there is anything morally right (or wrong) about seeking to propagate one's genes in any particular way. True, one could fit it into Aristotle's doctrine of the mean, whereby "Virtue, then, is a state of character concerned with choice, lying in a mean, i.e., the mean relative to us, this being determined by a rational principle, and by that principle by which the man of practical wisdom would determine it" (Aristotle, 1908, Bk. II.6). One could have the right amount of children, in the right way, at the right time, and so forth, with what is right being determined by whatever propagated one's genetic material best. But, it would not serve as a good basis for moral theory.

First, the essentialist in Aristotle could not accept this circumstance, since it is not unique to humans. All other beings, from amoebas to zebras, have the same function on this account.[5] More importantly, it fails to meet the additional criterion for virtue that he provides later in NE, Bk II.6, "Therefore, if this is true in every case, the virtue of man also will be the state of character which makes a man good and which makes him do his own work well." Unless one is willing to embrace a very strange form of the term "virtue," successful genetic propagation does not necessarily make one do one's own work well nor make one good. As an illustrative example, one can attain success at this function in multiple ways. Perhaps the most effective way of attaining this goal is by cloning, division, or another method of reproduction that preserves one's entire genetic makeup. Sexual reproduction is a distant but still important second-best method, as it preserves roughly half of an individual's genes, and carries other evolutionary benefits for the sexual reproducer (Barton and Charlesworth, 1998). But, there are other methods of ensuring that some of one's genetic material enters the next generation, such as acting to benefit siblings, cousins, and other blood relatives who carry at least some of the same genes as that individual.[6] Intention to propagate genetic material would not be relevant, nor would method, except insofar as one's method was more effective. Methods sometimes practiced by animals in the wild—and in human history—of rejecting or destroying children who do not share one's genetic material could, perhaps, be thought good. Only under a notion of morality at odds with most common conceptions of morality could this "function" serve as a virtue.

More importantly, perhaps, it would seem quite odd to connect morality to this function, as many beings that could not possibly have the capacity to act morally in any interesting sense would seem to have the same capacity for moral action. The same behavior for the same purpose (propagation of genetic material) could be performed by, say, sharks, chickens, and humans. Either all are similarly moral, or the action would be non-moral in sharks but moral in humans. It is perhaps possible to explain all of these concerns, but it is much more compelling to hold that if this is a function of humans—which is by no means proven—it is amoral.

From this, one might conclude that natural facts, absent a creator, have no moral import. Indeed, over the years many have concluded thus and denied Darwin on these grounds. Samuel Wilberforce, bishop of Oxford at the time of the publication of *The Origin of Species*, stated

> Such a notion is absolutely incompatible...with the whole representation of that moral and spiritual condition of man... [Multiple fundamental claims of Christianity] all are equally and utterly irreconcilable with the degrading notion of the brute origin of him who was created in the image of God...(Rachels, 1990, p. 47).

Others have argued similarly in the intervening century and a half. Wilberforce spoke of evolution as a threat to Christianity, but similar concerns may seem to apply to natural law generally. However, it does not follow that natural facts can have no moral importance absent an intentional function of humans. It does follow that these natural facts do not have the same meaning as is sought by natural law theorists; but natural facts can still have very important implications for moral thinking. There are specifically evolutionary ways of comprehending the moral import of moral facts; as well, appreciating the evolutionary origin of facts about human existence affects the recognition of what sorts of facts can matter morally and how ethical thinkers ought to consider them. There are at least three important moral functions of natural facts that exist in this context. I will make brief comments about two such functions, then discuss a third in depth.

3 Utility Monsters and other Trivia

Natural facts can serve as limits on the extent to which certain moral arguments ought to be pursued. This is not a commonly accepted claim, but it is compelling. If a moral argument depends upon circumstances which are not possible, or are outrageously unlikely, due to the natural facts of the world, that argument should be thought largely unimportant for determining answers to moral questions.

Such arguments are common in some philosophical endeavors. Consider the classic utility monster objection: if a sadist derives so much pleasure from torturing little children that the torture produces more good in him than it produces harm in the children and community, then utilitarianism obligates persons to turn over their children to be tortured, which is absurd. This objection depends upon the existence of a person who gains much more pleasure from harming others than is lost in the harm, and a good response to such an objection is to note that the natural construction of humans and their happiness makes it simply not possible, or at least fantastically improbable, that any such human could exist. Since there is no reason to think that any human being could possibly enjoy torturing another so much that this pleasure even outweighs the harms caused by it, much less that anyone could enjoy that torture so much that torturing another is the action that produces the most good for all, it is appropriate not to allow our moral reasoning to be guided by it.[7]

This is a topic deserving of further discussion, but I will not belabor the point here. It is important only to note that natural facts play an important role in this moral debate, and whether a given argument should be critical or marginal can be (and, I hold, should be) determined in at least some cases by whether the situation under consideration is possible, given the natural facts about humans and their environment.

4 The Origin of the Ethical Species

Historically, utilization of evolutionary theory in ethics has been fraught with errors, self-serving rationalizations, and willful misinterpretations of the theory ever since Darwin first proposed it (Singer, 2000; Dickens, 2000; Kerr, 2002).[8] The study of ethics has been erroneously reduced to equating "good" with "more evolved"—a claim which errs both in its understanding of ethics and its understanding of evolution. Ethics has also been reduced to sociobiology and to Skinnerian behavioralism, or thought to be rendered entirely meaningless, by various readers and mis-readers of Darwin and those who have followed him. Thus, to speak of evolution and ethics is to run the risk of being mistaken for any one of these erroneous views, especially as the exposition here will be brief. Yet, it is important to point out at least one important moral concept that is ensured by an understanding of evolution, and entails natural facts that should be recognized as morally relevant.

Evolution describes changes in species as being gradual, and leads one not to expect quantum leaps. Though significant differences between related species may exist in any given time period due to the intermediary species' current non-existence—though see discussion of "ring species" as interesting examples of the concurrent existence of intermediary species (Irwin et al., 2001a,b)—those changes will have happened gradually. Consequently, we can expect human beings to be similar to closely related species in many different ways. Moreover, we can understand "closely related species" to include quite a few species—including not just primates but also most, if not all, mammals.

A careful recognition of the nature of evolution indicates that many of the qualities we take to be morally salient in human-to-human interactions are also present in many human-to-non-human-animal interactions. We are neither physically nor psychologically unique, but rather different only in the extent of the development of various features we as a species have found useful in our environments. The natural facts of pain, capacity to suffer psychologically, social ties, and the like are present in greater or lesser extents in many beings other than humans. If these facts mean something in moral deliberations for humans (and, as will be seen below, I believe they do) then they must also be considered in moral deliberations about human treatment of non-humans. This does not, strictly speaking, show that the existence of these natural facts necessitates any moral relevance to the facts in question; rather, it entails that any moral relevance that these facts have in one scenario with one species (human-to-human interactions) must, unless successfully argued otherwise, have a similar impact on a relevantly similar scenario involving a different, non-human, species.

How they must be considered is a matter for significant debate, much of which has been done elsewhere (Dombrowski, 1997; DeGrazia, 1996; Singer, 1993; Carruthers, 1992; Regan and Singer, 1989; Regan, 1983). Consistency requires that the facts must be accounted for in some fashion. Any moral theory which fails to address, e.g., the facts of animal suffering, or which "deals with" such suffering *via* a Cartesian mechanism of denying its existence is significantly lacking. As well, a theory that deals with these similarities by denying the relevance of the same features in non-humans while granting them significance in humans must produce some valid, non-*ad-hoc* argument for that dichotomous treatment. This is not a simple task, as shown by authors such as Regan and Dombrowski (Dombrowski, 1997; Regan, 1983), though some have treated it as such. It is a valid objection to a theory that it inappropriately dichotomizes humans and non-humans with regard to characteristics they have in common; perhaps more importantly, a proper recognition of the multi-species existence of these natural facts would entail a significant change in many moral theories and positions.

5 Essentialism, Moral Acquaintances and Natural Facts

Finally, natural facts can be important for understanding moral interaction by defining features of human existence that are morally relevant. Such a pursuit abandons the search for a human function, as understood by natural law, but does seek important features of human existence. Let us return, for the moment, to Aristotle's development of moral theory; for if his analysis cannot define a necessary *telos* for man, it may be able to assist in explaining the relevance of natural facts about humans for moral development. Aristotle's view can be interpreted as an essentialist view where one describes, not a human function, but *a posteriori* the essential features of human existence. Martha Nussbaum has expanded upon such a view (Nussbaum, 1992). She argues for a "thick vague conception" of a human life, which includes such things as mortality, having a body, infancy, capacity for pleasure and pain, a social nature, need for humor and play, and so on (pp. 216–220). Any being missing these fundamental features would not be thought to be human—she mentions the Greek gods, who have human form and emotions, but neither mortality nor infant development (p. 216). She also argues that capabilities are a part of human existence, such as being able to be well nourished, to develop a theory of the good, to use the five senses, and so on. She claims that a life that lacks any one of these capabilities "is lacking in humanness" (p. 222), though presumably not entirely inhuman. The important point seems to be that these should be "a focus for concern, in asking how public policy can promote the good of human beings," such that a life that is lacking in these capabilities is not a good human life (pp. 221–222). Similarly, one could suggest that they are a relevant focus for moral concern.

One concern with this position that Nussbaum recognizes is that natural facts, absent the concept of a guiding function, do not automatically imply any moral import. Since she does believe that these natural facts have moral import, one might wonder whether Nussbaum has described a type of natural law theory, since both share similar sources (natural facts) and results (human goods). This concern is amplified when one notes that Nussbaum's list contains few surprises for many natural law theorists. Natural lawyers have composed catalogs of basic goods that resemble Nussbaum's: Gómez-Lobo discusses life, family, friendship, work, play, knowledge of beauty, knowledge, and integrity; Murphy includes life, knowledge, aesthetic experience, excellence in work and play, excellence in agency, inner peace, friendship and community, religion, and happiness, and Finnis, Grisez, and others provide similar lists (Murphy, 2001, 2002; Gómez-Lobo, 2002). So has she created a natural law theory through the method of essentialism?

She has not. Nussbaum's move from facts and capabilities to the good life for humans is not at all implausible, but it does not derive the good for humans from the *function* of humans. It seeks to derive the good for humans from human nature, as perceived by those living as humans. Her analysis draws no distinction between perceiving humans as the watch or as the stone. Whether we were created specifically to be the way we are, or whether we merely developed this way as a series of lucky coincidences, the fact is, we exist now in this form; how we got here is less important than our current facts of life and capabilities. Especially for the development of moral theory, what is important is what we *can* do, not what we are *intended* to do.

This distinction means that the essentialist still must explain the moral importance of those capabilities to derive moral claims from them. Whereas the appeal to human function provides a reasonably clear answer to why natural facts matter morally, the essentialist does not have that to which to appeal. To avoid a simple transition from is to ought, a rationale for the relevance of these characteristics must be provided.

6 Moving from an Is to an Ought

The objection that one may not derive an "is" from an "ought" does not seem to have much effect on a natural law theory. The objection is, roughly, that one cannot derive an "ought" statement directly from an "is" statement, because they are two distinct types of statement; one type will never produce the other. But on a natural law account, this is not exactly correct. Factual statements are still different from normative statements, but natural facts about the function of humanity are derived from the way humans are created. We are created in a particular way, because we were supposed to be created in that way. Deriving normative statements from factual statements about human nature, at least insofar as those facts reflect Aquinas's eternal law, the rational plan by which all creation is ordered, is not surprising. Those facts exist because they ought to exist; we are made in a certain way because that is the way we ought to be. For natural facts reflecting the eternal law in a natural law theory, the "is" of the fact statements derives from an eternal "ought." Deriving a normative statement back out of such a factual claim is just a matter of reverse engineering the facts back to the "oughts" that they began with.

If the factual account provided by Nussbaum cannot be described as a natural law theory, however, it cannot appeal to such a method of deriving normative meaning from the factual claims she discusses. If there is normative value to these facts themselves, it must be grounded in some different rationale. Given the rigorous opposition to the derivation of moral values from facts from Hume forward, this is a challenge; however, a legitimate rationale can be provided for the consideration of such natural human facts outside of natural law theory.

Such a rationale can be provided by the mechanism of "moral acquaintanceship." This concept, as developed by Erich Loewy, has important implications for the normativity of natural facts (Loewy, 1997). Loewy argues that there are several features that can be known about human nature, which he calls "existential *a prioris*." They are:

> (1) an inborn and necessary drive to survive ("being"); (2) the satisfaction of biological necessities (food, drink, shelter, etc.); (3) the meeting of social needs (different though they may be, all creatures have social needs); (4) a desire to avoid suffering; (5) a common sense of basic logic (at least sufficient so that the incompatibility of "p" and "non-p" are evident); and (6) a desire to lead their own lives in their own way (p. 3).

These *a prioris*, which can fairly be described as natural facts, allow him to derive a thin moral system that, though it does not describe the whole of human morality, sets some basic moral groundwork while defending some more specific moral claims (Loewy, 1997, Chapters 4–5). He does not, however, claim to derive an "ought" from an "is," arguing rather "that human experience and biology provide the necessary conditions for all else, including ethics. ... Saying that something ought to be because it is, is evidently wrong: but we cannot begin to think about what ought to be the case unless we are quite clear about what is the case" (p. 41). He does not hold that these facts, or even a larger set of facts including these and others, are in any way a *telos* for humanity. They are not definitive of morality or the good life; they merely serve as boundary lines and necessary conditions for what can be considered morality.[9]

Note, for example, some of the broad moral principles he derives from these *a prioris*:

> ...(1) an understanding that all of us are united by a common framework of needs, interests, and capacities; (2) the necessity of keeping the peace, maintaining mutual respect, and utilizing an agreed upon process for resolving problems; and (3) the desirability of assuring full access to the basic biological and social necessities we all have and all know we have (p. 163).

These principles are rather gauzy by most ethical theories' standards; though the second implies some straightforward right and wrong policies and actions, the first and third contain guidelines but do not easily lend themselves to moral imperatives. Though he does derive some very specific conclusions—for example, he argues that "the attempt to interfere with birth control is ethically indefensible" (p. 197)—for the most part the *a prioris* stand as guard rails for ethical thinking. No claim to be moral can fail to appreciate our common framework or the desirability of food, water and shelter, but there are many possible moral claims that do take these into account in rather different ways. His third principle, for example, states that the desirability of basic resources for humans is undeniable, but it leaves open to further development what obligations that fact entails. The *a prioris* are not natural laws that are definitive of ethics, but rather the preconditions for ethics (p. 106). On this account, many different possible moral theories can come out of these preconditions—one could consistently question whether the content of Loewy's own interpretation is a proper derivation from the existence of the *a prioris* and still agree that the natural facts of the *a prioris* are morally important. All different readings of morality that are consistent with these principles are allowed by these natural facts; but those that fail to meet these obligations are not.

In the same way that a potter must know the nature of the clay she is working with before she determines what to make out of it, ethical theorists must know the nature of the beings for which they are developing ethical theories. These facts about human beings serve as limits to moral theory, much as the factual features of musical instruments limit musical composition: a symphony requiring the piccolo section to play bass notes will not work well. But they serve not only as limits, but also as guides. As Singer suggests in a different context, a good woodcarver will examine a piece of wood before working it to follow its natural grain when carving a bowl (Singer, 2000, p. 40). The same should be true of ethical theorists, who should use these natural features as guides, though not as absolutely defining factors, in the development of moral theories.

Therefore, natural facts are crucially important to understanding the moral acquaintanceship that Loewy discusses, and important for moral theory in general, and this importance begins with the existence and moral relevance of these *a prioris*. As noted above, Loewy argues that these are features of the human existence; not natural or ethical laws, but preconditions for the development of ethics. They are natural facts of human existence. More importantly for the understanding of their normative value as natural facts, they form the basis for human existence, ethics, religion, society, and the whole of any human endeavor. Thus, natural facts, on this account, are crucial to ethics simply because they provide the necessary framework in which any ethical analysis is possible.

If this is true, then natural facts have normative value in defining the circumstances under which moral reasoning and development must function. The *a prioris* determine the preconditions for morality. The implication for Nussbaum's lengthier list of human characteristics is that one may understand that list as an expansion of the basic preconditions of human existence, understood both in terms of fundamental features of human existence as well as descriptions of human capabilities (and limitations.) These, too, underlie the practice and development of moral theory, and place limits upon it. This may appear to a natural law theorist to be a very limited—almost non-existent—moral role for these natural facts, but, as shown below, they can lead to very concrete and important moral claims.

7 The Importance of Contingent Facts

As suggested in the challenge to Nussbaum above, one might also ask whether Loewy is not just introducing a natural law account by the back door. Such an interpretation can be read into some of Loewy's language when he calls the existential *a prioris* "the 'givens' of our existence" and notes that they are "rooted in the human condition" (Loewy, 1997, p. 105), but it would be an error to do so. He immediately counters his own language by his direct refutation of the claim that he is proposing a system of natural law (p. 106). Of course, persons may claim to be doing one thing, and even believe it, while still actually doing another,[10] so his explicit denial may not serve as sufficient proof that this is not a natural law theory. That proof must come from the details of the theory itself.

Loewy's moral acquaintanceship account differs from a natural law account in at least two important ways. First, rather than developing a *telos* of human existence, and deriving moral meanings from that, Loewy seeks only to understand the necessary underpinnings of human existence *qua* human existence as we know and experience it. Many possible different and even conflicting (individually chosen) functions can and do exist within the scope of Loewy's moral acquaintanceship defined by these existential *a prioris*. His goal is, therefore, not only more limited than that of a natural lawyer, but it is also different in scope. He seeks not to find a natural human *telos* but to circumscribe certain limits of possible chosen goals and behaviors by the necessary features of human existence.

The second important difference between Loewy's *a prioris* and a system of natural law is that Loewy's system recognizes that purely contingent facts can have great moral import in a way that is not possible for natural law. Natural law theories traditionally seek to locate the human interaction with eternal law, or a consistent *telos* for humanity in general. Natural laws are set in place by nature and we discover them. Neither human invention nor chance happenstance can alter these laws, nor create another. In contrast, the moral acquaintanceship that Loewy finds in humanity is created by contingent facts, and can change as those contingent facts change. The very notion of a *telos* as a set, designed function does not allow for facts which are one way, but just as easily could be another, to have moral importance. If natural law is, per Aquinas, how humans partake in the eternal and unchanging law, then fundamental contingencies cannot be a part of that natural law (*ST*, IaIIae 91, 2). The account of moral acquaintances sketched herein allows for them to be as relevant as any unchanging fact of nature.[11]

A good example of the importance of purely contingent facts in the moral acquaintanceships that Loewy develops can be seen in his approach to birth control. Contrary to many natural law theories, he argues that the prohibition of artificial means of birth control is ethically indefensible. This decision is motivated by contingent natural facts, in combination with his *a prioris*. It is a fact that many people are malnourished and even starving to death in some of the most overpopulated areas of the world. While it is clear that allowing or even promoting birth control would not itself eliminate these problems, he finds it unacceptable that any position could hold birth control to be immoral when such a position contributes to this sort of suffering. His embrace of artificial birth control methods is due not to some notion of human flourishing that involves birth control, but rather due to the natural facts of the matter in the world today.

I do not mean to minimize the importance of contingent facts of the case for natural law theories. All moral systems use natural facts in this way to define cases: were a moral theory to hold it to be good to feed the hungry, one would still need to understand the contingent facts of a given case to determine whether a person was hungry. (Has he eaten recently? How much?) The importance of contingent facts on the view outlined here is stronger. The contingent facts of the context in which the prohibition on birth control is being rejected serve as the entire grounds for Loewy's rejection of it. He argues,

> [I]f populations continue to grow at the present rate, nourishing the earth's population will prove to be an impossible task... Unless we choose not to treat or to prevent disease (not really a palatable thought!), controlling birth rates seems to be the only viable conclusion(Loewy, 1997, p. 196).

It is not important here whether or not Loewy's Malthusian predictions are correct, or timely. What matters for the understanding of the importance of natural facts is that they serve for him as the grounds for the wrongness of restricting birth control. Were the facts different, his argument as it stands would not exist, and his conclusion would not follow.[12] This can be seen by changing those facts to a different, hypothetical case.

Loewy discusses the appropriate attitude towards contraception "[s]tanding outside a poverty [stricken] area [teeming] with children some of whom are sold and almost all of whom are homeless and starving..." (p. 197). Under those circumstances, he argues, if compassionate rationality guides our decisions, and if existence has any moral relevance at all, then opposing birth control is ethically indefensible (p. 197). However, consider the same features (compassionate rationality and existence having moral relevance) under very different circumstances: Margaret Atwood's world of *The Handmaid's Tale* (Atwood, 1986). In the near-future world of this novel, dramatic political upheaval has led to the replacement of the United States government with a theocratic regime called the Republic of Gilead. At the same time, chemical spills and pollution have lead to dramatically lower fertility rates, such that the few women that are fertile are regarded as a resource to be hoarded by the more powerful men in society. Though a consideration of compassionate rationality would surely argue against virtually all of the features of the Republic of Gilead's brutal and depersonalizing totalitarian society—and Loewy would emphatically reject such a society—the fact of decreased fertility would have a powerful impact on Loewy's argument against restricting or prohibiting contraception. Instead of persons' inability to access contraception leading to the birth of further persons who will suffer greatly, any children born in these circumstances will be valued highly and represent the best hope for the further existence of any human beings. The importance of existence and compassionate rationality might argue even in favor of restricting contraception in this case; but the important point is that they would not contribute to an argument against such restrictions. The natural facts in this case do not determine whether a particular moral guide applies in this case; they determine whether the moral guide exists.

One might reply that the contingent facts in the above case do not actually determine the existence of the moral guide that drives the argument, but merely serve as the "facts on the ground" which help to understand it. On this line of argument, the moral guides are the existential *a prioris*, from which the importance of being and the meaning of "compassionate rationality" are derived (Loewy, 1997, pp. 120–123). These remain constant regardless of the surrounding factual context; thus, the basic moral guides remain the same, while the facts of the situation vary.

In other words, one might argue, this is nothing more than examining the contingent facts to determine whether and how stable rules apply to a particular case, which, as noted above, natural law moral systems do. If so, then contingent facts are not relevant to moral principles, but only necessary facts. Thus, Loewy's argument ends up looking much more like a natural law theory, wherein the natural facts are necessary and immutable. Perhaps that initial resemblance could support an argument that Loewy has only named some but not all of the *a prioris* of ethics; and, were he to name them all, he would have developed a natural law theory. If so, this supposed opposing theory is actually a natural law theory.

There are two responses to this account. The first is that the natural facts in the above case do not function as means of determining how a case corresponds to a given rule, or to which rule it corresponds. Rather, the facts determine the actual content of the rule. The natural facts determine whether the prohibition of birth control is "ultimately evil," as Loewy charges is the case in modern circumstances (p. 197), or whether it is morally neutral or perhaps even preferred over making it widely available, as might be argued in Atwood's possible future. The facts of the matter define the content of the rule about birth control, not just whether a given case applies.

In reply, one might argue that the rule about birth control was not a moral rule, but rather an application of the more general moral rules of, "show compassion" and "respect the moral importance of 'being,'" as developed from the existential *a prioris*. In this case, the moral rules remain the same but how one shows compassion and lives up to the requirements of the *a prioris* differs. Again, the argument would be that the contingent facts are relevant here only in the same way as they are in any other theory, and Loewy is merely reflecting, perhaps only partially, a universal moral truth.

However, such a line of argument would be ultimately counter-productive for defenders of natural law, as it tends to undermine the very nature of the rules natural law theory itself seeks to define. The same sort of argument could be applied to the moral rule *against* birth control in many theories of natural law. That moral rule derives from the arguably necessary connection between sexual intercourse and procreation as well as its ability to increase love and bonding within a marriage. Any separation of "the unitive and the procreative" would fail to "preserve in its fullness the sense of true marital love and its orientation toward [humanity's] exalted vocation to parenthood" (Paul VI, 1968, Section 12). Such a description of sex is only possible given the nature of sex in the human species.[13] Were human physical and social society only slightly different, the unitive aspect of human sex would also be different.

Whether or not it is procreative, the unitive nature of human sexuality is generally defined by units of two persons engaging in sex that unites them more closely. Sexual intercourse outside of that pairing often fractures the partner bond; successful sexual bonds of more than two humans in one group are generally rare. There are, of course, many exceptions to this general rule, but it is at least arguable that this is the fashion in which human sexual relations unites individuals most effectively.[14] Were humans more like our near cousins, the bonobos, who use sexual contact and intercourse of all sorts as a kind of universal social glue, this might not be the case. According to de Waal and Lanting, bonobos use sexual contact, whether male-female, male-male, or female-female, in groups of two or more, in many different fashions and positions, as a means of overcoming social tensions (de Waal and Lanting, 1997, Chapter 4). Situations which might in other primates be resolved through violence or the threat of violence tend to be resolved by bonobos via sexual contact. Sex is used "to promote sharing, to negotiate favors, to smooth ruffled feathers, and to make up after fights..." (de Waal and Lanting, 1997, p. 112). Female-female sex also serves to assist in the integration of younger females into existing groups, and female bonds for bonobos are an important part of their social structure. Obviously, many of these contacts are not at all procreative in nature; even some of the male-female contacts are brief and do not appear to end in orgasm.

Human sexual capacity is physically quite similar to that of the bonobo. Bonobos also engage in sexual activity apparently simply for pleasure, including oral stimulation and masturbatory behavior in males and females; there is physiological evidence of female orgasm as well. Females also develop genital swellings (which in chimpanzees and bonobos indicate increased sexual receptivity and are sexually attractive to males) well before their cycle makes them fertile, and also when they have given birth to an infant too recently to be fertile again yet. Like the bonobo, and unlike many mammals with "heat" cycles, humans are capable of sexual activity for pleasure, in non-procreative fashions, and when procreation is not physically possible.

The point of the above is to indicate the importance of sexual contact of many different forms as a unitive part of the bonobo's entire social structure, not just between procreative partners. The natural function of sex for bonobos is not one which combines the procreative and the unitive between two partners, but rather one that serves as the tie which, spread promiscuously throughout the social group, binds them all together. Humans did not develop that way, but could have, as we share with bonobos the ability to engage in and enjoy sex in many times and ways other than when and how it is most likely to be procreative.

Consequently, if the above argument against the interpretation of Loewy (that the specific rule about birth control is merely an expression of the more general rules of "respect 'being'" and "exercise compassionate rationality") is accepted, it seems the same response must apply to the Vatican's (and many others') law against the use of birth control. That law depends upon the nature of human sexuality, but could be described as reflecting a more general moral rule, such as "preserve social stability." For it is merely a contingent fact that we are so constituted that sex for us, whether or not it is potentially procreative, is unitive largely only when performed between two partners seeking to bond themselves together. Were sex, for us, to serve the socially unifying function that it does for bonobos, our society would not survive if sex were restricted to heterosexual, married couples. The natural law prohibition against birth control (as well as those against homosexual and non-monogamous relations) would not follow from "preserve social stability," for it is due to the contingent fact that we are so constituted that single-pair partnerships are how we tend to bond sexually that such a rule could exist.

This would radically change the very nature of the Catholic Church's position on birth control and homosexual intercourse, even if the implications of the rule remained the same. The Vatican's argument against non-procreative forms of sexual intercourse cannot be reduced to a more general law without destroying the argument, as it depends upon the nature of sexuality, not upon preservation of social stability. In the same way, one cannot reduce Loewy's rejection of arguments against birth control to expressions of a more general, universal law without destroying the very nature of the argument. The contingent facts about birth and birth control in the world in which we live determine the rule, but the nature of the rule is not merely as reflection of a higher duty. Therefore, it should be concluded that contingent facts determine Loewy's moral rule about birth control, and it cannot be meaningfully reduced to a universal norm. Contingent facts are morally important for his theory.

There is also a second reason why Loewy's system must be understood as allowing contingent facts to define moral rules in a fashion quite apart from natural law: the moral *a prioris* themselves depend upon contingent facts. Even though Loewy argues that the *a prioris* are the necessary framework in which any ethics can be developed (Loewy, 1997, p. 4), and a "fact of sentient existence" (p. 106) that we "inevitably share" (p. 141), they are in a very important fashion, contingent. They may be "the obvious foundations of our own existence" (p. 106), but they, and that existence for which they are the foundations, are contingent upon the type of beings that we are. The *a prioris* are not necessary in the sense of being laws of nature; they merely are. Loewy argues for no more importance than that; yet that makes them fundamentally important for any human endeavor, since they underlie everything that we do.

Loewy argues that they are necessary as a framework for human endeavor; he claims that they necessarily apply to all sentient beings and strongly suggests that they apply to all possible sentient beings (p. 105). The latter claim is not quite correct. Some of Loewy's *a prioris* may be necessary features of all sentient beings, but different facts about different beings can make them radically different in meaning, such that it would be appropriate to describe the *a priori* as a different claim. For example, his fourth *a priori* is "a desire to avoid suffering." What this means differs due to contingent facts about particular persons and beings. Persons with the rare genetic condition of hereditary sensory and autonomic neuropathy [HSAN] feel no physical pain. This is not usually a boon for these persons. Most versions of the mutation carry with it many clearly negative additional effects, but even for those who suffer few other side effects, this condition is not generally beneficial: most forms appear in infancy and are accompanied by self-mutilation, inability to regulate body temperature, and inability to detect and respond to dangerous and normally painful stimuli (Weingarten et al., 2006; Low 2002; Haworth et al., 1998). However, any desire that such persons have to avoid suffering has no connection whatsoever to physical pain. In some types of this disorder, due to a selective absence of the nerve fibers "which are important for sensing the sharp, well-localized, and prickling sensations of pain" these persons are not only incapable of feeling pain, they are actually physically incapable of even the physiological precursors to feeling physical pain (Weingarten et al., 2006, p. 339; Low, 2002, p. 607).

Of course, such persons are extremely rare carriers of what is largely a dangerous genetic variation from the human norm, and consequently do not provide a strong argument against those who would argue that there is a fundamental human nature that includes the capacity for physical suffering. Two points, however, are important to note here. The first is that it is a contingent fact that we feel pain as suffering, though not an unexpected one. The mutations that lead to HSAN are rare, and for good reason. The self-destructive behavior associated with such mutations alone would historically have made those who carry active forms of the condition from birth very unlikely to survive to procreate; combine that with the inability to respond to negative stimuli and it is doubtful any would have survived for long until very recently. But they do exist, and they illustrate an interesting point. It is easy to understand the evolutionary value of being able to respond to negative stimuli, which is likely why virtually all complex animal species seem to have at least some capacity for such response.[15] But it is not necessary, as HSAN mutations show. Also, there is no particular reason why the response to these stimuli, especially in beings with higher mental functions, must entail suffering. Since the actual perception of pain stimuli as the very unpleasant sensation of pain occurs when the stimuli are interpreted by the brain, there is in principle no reason why we could not experience "pain" stimuli as a tickling sensation, the sensation of being rubbed with a piece of fur, or, for that matter, the sensation of hearing an A over high C played on a kazoo. In practice, of course, there are good reasons why sensing pain as something extremely unpleasant was the most likely evolutionary pathway for these sensations to follow, but the fact that it did is a contingent (though quite understandable) fact.[16]

More to the point, however, is the idea that humans could soon be in a position to alter the way that they experience pain. Some forms of HSAN occur because of a lack of or malfunction in pain receptor nerves; however, some forms do not appear to produce any abnormality in the nerves. Rather, the indifference to pain may be related to abnormalities in pain processing or connectivity between properly functioning pain nerves and the brain (Weingarten et al., 2006, p. 339). If we can understand how these mutations function, in coming generations we may be able to alter the sensation of pain (at least in adults) to render it as experienced entirely without suffering. Were this to be possible, and were it to become widespread, Loewy's understanding of the relevance of physical suffering would change apace, and it could even eventually be recognized as irrelevant due to the very contingent fact of its ceasing to be suffering. In any case, it would not be a universal feature of human existence. Contingent facts make his *a prioris* what they are; different contingent facts would make them differ.

This is not a great difficulty for Loewy's system, though; the *a prioris* are "existential." That is to say, they are "conditions dealing with existence and known by experience.. . .[They are] known, and taken for granted, about each other from everyday common human experience" (Loewy, 1997, pp. 104–105). Counterintuitively, his *a prioris* are known *a posteriori*. Therefore, if the basic conditions of human existence were to change, the *a prioris* could change along with them. Such an approach would be much less compatible with some other views of human nature, such as that represented in John Paul II's "Apostolic Letter Salvifici Doloris..." in which he notes the "salvific nature" of suffering (John Paul II, 1984). He notes in Section 2 that "what we express by the word 'suffering' seems to be particularly *essential to the nature of man*." He discusses many types of suffering in this tract and focuses on what he calls "moral suffering," but physical pain figures significantly in his analysis: "In fact one cannot deny that moral sufferings have a 'physical' or somatic element, and that they are often reflected in the state of the entire organism" (Section 6) and "It is obvious that pain, especially physical pain, is widespread in the animal world. But only the suffering human being knows that he is suffering and wonders why..." (Section 9). Were physical suffering to cease to exist, defenders of such a position would have to address a fundamental change in something taken to be a crucial feature of human function and existence. A theory like that presented here could simply recognize the change in the contingent facts that matter and move on with the new set of existentially discovered facts about human existence.

More generally, the basic principles that he derives from his *a prioris* are dependent upon contingent facts about human beings and the world in which they live. David Hume gives an instructive guide in his analysis of the source of the moral concept of justice (Hume, 1983, Section III, Part 1). Were the world constructed so that there was such abundance of all external needs that persons could, without any planning or forethought, easily obtain all that they needed for both survival and luxury, then the concepts of justice and private property would not exist. More explicitly for the purposes of the discussion here, Loewy's basic principles of

> ...(2) the necessity of keeping the peace, maintaining mutual respect, and utilizing an agreed upon process for resolving problems; and (3) the desirability of assuring full access to the basic biological and social necessities we all have and all know we have...(Loewy, 1997, p. 163)

would more or less cease to be relevant. If all persons could easily obtain full access to all biological and social needs, there would be no need to focus on or consider as important any mention of that desirability. It also seems likely that there would be less need to maintain the peace or resolve problems, and perhaps none at all, if everyone were able to keep themselves easily in whatever fashion they desired. The world is not constructed in this way, of course, and consequently these basic principles are critically important. But it could be otherwise.

8 Conclusion

It has been shown that the mechanism of describing the moral importance of natural facts in natural law theories can be dealt a blow by an understanding of the nature of an evolutionary development of humans. The concept of an intentional function of human existence that serves a vital role in natural law theories is neither entailed by nor necessary in an evolutionary picture; consequently, arguments about human goods that depend upon there being a set and intended human nature need not apply. However, it remains true that there are common features of human existence that can be identified, regardless of their origin. Through Loewy's account of moral acquaintanceships, these natural facts can be understood to have moral significance great enough to provide some moral guidance. Many moral views may be compatible with the significance of these facts, but at least some directive moral statements can be made based solely on the normative value of natural facts.

Notes

1. Which is not to take a position one way or another about whether religion and evolution are compatible; I leave that discussion for elsewhere.
2. According to Gould, "Humans are here by the luck of the draw, not the inevitability of life's direction or evolution's mechanism" (1997, p. 175). Humans, or even just complex beings like humans, are not inevitable.
3. This human feature is employed in Rorschach tests, which seek to determine how persons organize their perceptions of random ink blots. A more amusing example can be seen in Stephen Levy's "Does Your iPod Play Favorites?" (2006).
4. On this account, it is not clear that the individual being is actually the unit of selection or evolution. Dawkins argues convincingly in *The Selfish Gene* that the unit of selection is the gene itself, and a fair amount of study in the past three decades has tended to support this view, broadly understood (Dawkins, 2006). If this were the case, then perhaps this function would be that of the gene, not the individual human; if so, Aristotle's argument will not work.
5. One could quibble about this claim, arguing that the function of oaks is to propagate oak-genes, lions to propagate lion-genes, and the function of humans is to promote human-genes, thus providing a distinct, if similar, function for each. However, this derives from a misunderstanding of what this function is. An individual lion has no evolutionary interest in propagating lion-genes, *per se*. Rather, that individual is aimed towards propagating its genes; they happen to be lion genes, but the survival of the species itself is not a part of the individual being's function. The function of each being on this account, whether it be lion, llama, or loganberry, can be properly described as "propagation of that being's genes into the next generation."
6. This fact is often missed by persons who argue, e.g., that homosexuality has no genetic basis, since as it interferes with sexual reproduction, it would be an evolutionary "dead end" that could serve no evolutionary purpose.
7. For a similar example, see R.M. Hare "What is Wrong with Slavery?" (1986). Hare argues that even when a society theoretically appears to do much better when it is a slave society, he still is doubtful for reasons similar to those above that it really promotes the most good. Even if slavery makes a society better off than it was before, it seems highly improbable that the society could not be made even better off by eliminating the slavery; he also argues that there are few if any valid historical examples where it is even plausible that slavery could make a society better, especially considering its psychological effects.
8. Singer notes that only two months after the publication of *The Origin of Species*, Darwin wrote of a newspaper review which held that his work "proved might is right." See Darwin to Lyell (Darwin, 1887, p. 62; Singer, 2000, p. 10).
9. In what follows, I posit Loewy's methods as a means of understanding how natural facts matter, and I employ some of his conclusions to explicate how those methods work and how they differ from natural law. I do not expressly defend or reject the specific moral claims that he derives from those *a prioris*, as that is outside the scope of this work.

10. See, e.g., Susan Moller Okin's feminist critique of Rawls's *Theory of Justice*, where she argues that he proposes a theory that disempowers women by empowering heads of families, even though his basic principles should make gender irrelevant (1987).
11. Loewy does not carefully discuss this point, as this is not his goal; but it is developed here.
12. He might still argue that most arguments in favor of calling birth control immoral are, in his words, "a-rational" (p. 196), but his positive argument for allowing it would not function.
13. Given the fact that a fertile woman is only capable of becoming pregnant 2–3 days out of a cycle, and given the fact that sexual intercourse is possible and pleasant on other days, and that sexual intercourse is possible after menopause or while one partner is temporarily infertile, it is a bit odd to argue that the unitive and procreative are united in an "inseparable connection." Given these facts, it seems more appropriate to say that the unitive and procreative aspects of sexual intercourse are united somewhere less than 3 times out of 28. This matter is beyond the scope of this essay.
14. I am aware of the existence of societies where multiple wives are allowed or even encouraged, such as "fundamentalist Mormon" sects in rural Utah, or some cultures in the Middle East and Northern Africa. I am not convinced that such relationships actually tend to produce good, unifying relationships, but I need not argue this here. For the purposes of this argument, I will assume that Pope Paul VI's analysis of human sexuality is accurate; if it is not, the theory he espouses would be further challenged in its ability to prove these claims but the argument herein would still follow.
15. That mammals and birds have a similar capacity for physical pain as do humans is strongly implied by the relevant similarities in their nervous systems and physiological brain functions (DeGrazia, 1996, pp. 108–112). DeGrazia also notes that some evidence of the capacity for pain exists even in earthworms in that they appear to have the some of the same brain chemicals (endorphins) that humans use for pain stimulus and response.
16. It turns out to be difficult to give a clear definition of exactly what pain is. DeGrazia's given definition, "an unpleasant or aversive sensory experience typically associated with actual or potential tissue damage" (p. 107) seems to lack a clear understanding of the true suffering that physical pain brings. Fingernails on a chalkboard can be unpleasant; physical pain can be unbearable agony. A further exploration of this topic, though valuable, is outside the scope of this work, but DeGrazia gives a healthy discussion and bibliography of this topic (DeGrazia, 1996, Chapter 5).

Bibliography

Ackerman, D. (2004, June 15). I sing the body's pattern recognition machine. *The New York Times*. [On-line]. Available: www.nytimes.com/2004/06/15/science/15patt.html.

Aquinas, T. (2006) *The summa theologica of St. Thomas Aquinas* (Second and Revised Edition, 1920), Fathers of the English Dominican Province (trans.). [On-line]. Available: http://www.newadvent.org/summa.

Aristotle (1908). *Nicomachean ethics*, W. D. Ross (trans.). Oxford: Clarendon Press.

Atwood, M. (1986). *The handmaid's tale*. Boston: Houghton Mifflin.

Barton, N.H. & Charlesworth, B. (1998). Why sex and recombination? *Science*, 281, 1986–1990.

Carruthers, P. (1992). *The animals issue: Moral theory in practice*. New York: Cambridge University Press.

Darwin, C. (1887). Darwin to Lyell, in F. Darwin (ed.), *The life and letters of Charles Darwin*, Vol. ii. London: Murray.

Dawkins, R. (1996). *The blind watchmaker*. New York: W.W. Norton.

Dawkins, R. (2006). *The selfish gene*, 3rd edition. New York: Oxford University Press.

DeGrazia, D. (1996). *Taking animals seriously: Mental life and moral status*. New York: Cambridge University Press.

de Waal, F. & Lanting, F. (1997). *Bonobo: The forgotten ape*. Los Angeles: University of California Press.

Dickens, P. (2000). *Social Darwinism: Linking evolutionary thought to social theory*. Philadelphia: Open University Press.
Dombrowski, D.A. (1997). *Babies and beasts: The argument from marginal cases*. Chicago: University of Illinois Press.
Gómez-Lobo, A. (2002). *Morality and the human goods: An introduction to natural law ethics*. Washington DC: Georgetown University Press.
Gould, S.J. (1981). *The mismeasure of man*. New York: Norton.
Gould, S.J. (1997). *Full house: The spread of excellence from Plato to Darwin*. New York: Three Rivers Press.
Hare, R.M. (1986). What is wrong with slavery? in P. Singer (ed.), *Applied ethics* (pp. 165–184). New York: Oxford University Press.
Haworth, A.E., Thomas, N.H., Cook, L.J., Ellison, D.W., & Walker, J. (1998). Hereditary sensory and autonomic neuropathy with anhidrosis (type IV). *Journal of the Royal Society of Medicine*, 91(2), 84–6.
Hume, D. (1983). *An enquiry concerning the principles of morals*. J.B. Schneewind (ed.). Indianapolis: Hackett Press.
Irwin, D.E., Bensch, S., & Price, T.D. (2001a). Speciation in a ring. *Nature*, 409(6818), 333–337.
Irwin, D.E., Irwin, J.H., & Price, T.D. (2001b). Ring species as bridges between microevolution and speciation. *Genetica*, 112–113, 223–243.
Kerr, P. (2002, June). Saved from extinction: Evolutionary theorising, politics and the state.*British Journal of Politics and International Relations*, 4(2), 330–358.
Levy, S. (2006, January 31). Does your iPod play favorites? *Newsweek*. [On-line]. Available: http://www.msnbc.msn.com/id/6854309/site/newsweek.
Loewy, E. (1997). *Moral strangers, moral acquaintance, and moral friends*. New York: State University of New York Press.
Low, P.A. (2002). Autonomic neuropathies. *Current Opinion in Neurology*, 15(5), 605–609.
Murphy, M.C. (2001). *Natural law and practical rationality*. Cambridge: Cambridge University Press.
Murphy, M. (2002). The natural law tradition in ethics. *Stanford encyclopedia of philosophy*. [On-line]. Available: http://plato.stanford.edu/entries/natural-law-ethics/#1.
Nussbaum, M. (1992). Human functioning and social justice: In defense of Aristotelian essentialism. *Political Theory*, 20(2), 202–246.
Okin, S.M. (1987). Justice and gender. *Philosophy and Public Affairs*, 16(1), 42–72.
Paley, W. (1802). *Natural theologies*. [On-line]. Available: home.att.net/~p.caimi/oremia.html.
Plato (2006). *The republic*. R. E. Allen (trans.). New Haven: Yale University Press.
Pope John Paul II. (1984). *Apostolic letter Salvifici Doloris of the Supreme Pontiff John Paul II to the bishops, to the priests, to the religious families and to the faithful of the Catholic Church on the Christian meaning of human suffering*. [On-line]. Available: www.vatican.va/ holy_father/ john_paul_ii/ apost_letters/ documents/ hf_jp-ii_apl_11021984_ salvifici-doloris_en.html.
Pope Paul VI. (1968). *Humanae vitae*. [On-line]. Available: www.vatican.va/holy_father/paul_vi/ encyclicals/documents/hf_p-vi_enc_25071968_humanae-vitae_en.html.
Rachels, J. (1990). *Created from animals: The moral implications of Darwinism*. Oxford: Oxford University Press.
Regan, T. (1983). *The case for animal rights*. Berkeley: University of California Press.
Regan, T., & Singer, P. (eds.) (1989). Animal rights and human obligations, second edition. Englewood Cliffs: Prentice Hall, Inc.
Singer, P. (1993). *Practical ethics*, 2nd edition. New York: Cambridge University Press.
Singer, P. (2000). *A Darwinian left: Politics, evolution, and cooperation*. New Haven: Yale University Press.
Weingarten, T.N., Sprung, J., Ackerman, J.D., Bojanic, K., Watson, J.C., & Dyck, P.J. (2006). Anesthesia and patients with congenital hyposensitivity to pain. *Anesthesiology*, 105(2), 338–345.

Index